Approaches to Modeling of Friction and Wear

F.F. Ling C.H.T. Pan

Approaches to Modeling of Friction and Wear

Proceedings of the Workshop on the Use of
Surface Deformation Models to Predict Tribology Behavior
Columbia University in the City of New York
December 17–19, 1986

With 59 Illustrations

Springer-Verlag
New York Berlin Heidelberg
London Paris Tokyo

F.F. Ling
Department of Mechanical Engineering,
Aeronautical Engineering and Mechanics
Rensselaer Polytechnic Institute
Troy, NY 12180-3590
and Columbia University
Department of Mechanical Engineering
New York, NY 10027
USA

C.H.T. Pan
Department of Mechanical Engineering
Columbia University
New York, NY 10027
USA

Library of Congress Cataloging-in-Publication Data
Workshop on the Use of Surface Deformation Models
to Predict Tribology Behavior (1986 : Columbia
University)
Approaches to modeling of friction and wear.
1. Tribology–Mathematical models–Congresses.
2. Surfaces (Technology)–Congresses. I. Ling,
Frederick F. (Frederick Fongsun).
II. Pan, C. H. T. (Coda H. T.) III. Title.
TJ1075.A2W67 1986 621.8′9 87–28386

Camera-ready copy prepared by the authors.
Printed and bound by Arcata Graphics/Halliday, West Hanover, Massachusetts.
Printed in the United States of America.

9 8 7 6 5 4 3 2 1

ISBN 0-387-96656-0 Springer-Verlag New York Berlin Heidelberg
ISBN 3-540-96656-0 Springer-Verlag Berlin Heidelberg New York

PREFACE

At the conclusion of the Conference on Tribology in the area of Wear Life Prediction of Mechanical Components, which was held at the General Motors Research Laboratories and sponsored by the Industrial Research Institute, a very high priority recommendation was modeling of tribological systems. Since the appearance of the Conference Proceedings in 1985, the writers discussed the matter of modeling with Dr. Edward A. Saibel, Chief of the Solid Mechanics Branch, Engineering Sciences Division, U.S. Army Research Office.

This discussion led to a proposal for the Workshop which resulted in this volume. The choice of proposal and Workshop name turned out to be more restricted than it needed to be. As such, the Workshop adopted the name for this volume, *Approaches to Modeling of Friction and Wear.*

By design, the attendance was restricted to not more than 40 individuals so as to allow small group discussions. There were four panels which deliberated on the same questions after two invited area lectures. Section 1 contains the substance of the two lectures. Section 2 is the Workshop Summary which is a distillation of the four panel reports by the entire Workshop attendance. This was formally written up and edited by the eight panel session chairmen, i.e., each of the four panels met twice on two different questions under the leadership of a chairman for each session.

Section 3 contains four brief position papers on the subject of the Workshop. Section 4 contains twenty brief scientific papers which is an innovation of this Workshop. That is, attendees were offered the opportunity to write brief papers on modeling of friction and wear. It is gratifying to see the high percentage of attendees who did respond to the challenge. Together with Section 3, there are twenty-four inputs.

By way of qualifications, the Workshop addressed only sliding wear, leaving out other forms of wear such as cavitation and erosion. Although friction appears in the title, the Workshop did not discuss it at length. Future attention should be paid to mild wear and to the connection between micro-models and macro-models, the link between chemical models and mechanical models.

We hope this volume will be a vehicle for the Conferees to share their thoughts and ideas with the larger community and interested parties and we are honoring Dr. Saibel with this volume. We are, of course, delighted that Springer-Verlag has shown interest in publishing the volume.

F. F. Ling
Troy and New York, New York

C. H. T. Pan
New York, New York

EDWARD A. SAIBEL

Edward Aaron Saibel was born December 25, 1903 in Boston, Massachusetts. He attended the Boston Latin School. Ed received a B.A. degree in mathematics from the Massachusetts Institute of Technology in 1924 and a Ph.D. degree from the same institution in 1928. During 1924-1926, he studied at Cornell University. On October 3, 1986, he retired from the U.S. Army Research office where he had been Chief, Solid Mechanics Branch since 1972. He was adjunct professor at the Duke University, the North Carolina State University and the Wayne State University.

Ed started his teaching career at the University of Minnesota. It was there that he discovered his great facility as a teacher; he also discovered the former Lillian Howe who was destined to be his lifetime companion.

Ed and Lil have two children. Patsy is married to Dr. Rolf Winter and lives in Williamsburg, Virginia where both teach at William and Mary College. Mahlon lives in Cherry Hill, New Jersey with his wife Janice and works for Gerber Scientific. The Saibels have five grandchildren and two great grandchildren.

Between 1931 and 1957, Ed established a long association with the Carnegie Institute of Technology and the City of Pittsburgh as teacher, scholar, football coach, tennis coach, friend, mentor, violin virtuoso. Later he was to return during 1967-1972 as Institute Professor when Carnegie Tech was subsumed under the Carnegie-Mellon University.

During World War II, Ed was a civilian with the Office of Scientific Research and Development, U.S. Army. It was at this time he started his research in mechanics.

Between 1957 and 1967, Ed was Professor and, for the last seven years, Chairman of the Department of Mechanics at Rensselaer Polytechnic Institute.

Ed has had many students and proteges, and has authored or coauthored some 100 scientific and technical papers. With W. M. Lai, he coauthored the text, *Elements of Mechanics of Elastic Solids*. His research interests include topology, vibrations, stability, structures, mechanical properties, flow and fracture of materials, theory of lubrication, machining of metals and tribology.

He is a member of the American Mathematical Society, American Society of Lubrication Engineers (Fellow; Honorary Member; ASLE National Award), American Society of Mechanical Engineers (Fellow; Life Member; Mayo D. Hersey Award), British Society of Rheology, Mathematical Association of America, Society of Engineering Science (President, 1982-1983), Society of Natural Philosophy.

ACKNOWLEDGEMENTS

The success of this Workshop was made possible by the contributions of many people too numerous to mention here. We would like to acknowledge the contributions made by members of the organization committee who helped in many and often diverse ways. Particular thanks are due to Mrs. M. Sussman and Ms. J. Venable of Columbia University for their enthusiastic assistance throughout the Workshop and its preparation. We are indebted to Ms. J. E. Doocey of Rensselaer Polytechnic Institute for the preparation of the master laser-printed copy of the manuscript.

Acknowledgements are due to our sponsors: Columbia University and the U.S. Army Research Office under Contract No. DAAL03-86-G-0210.

TABLE OF CONTENTS

CONTRIBUTORS

Calabrese, Salvadore J.
Department of Mechanical Engineering,
Aeronautical Engineering and Mechanics
Rensselaer Polytechnic Institute
Troy, NY 12180-3590, USA

Chang, Yu Jun
University of Virginia
Charlottesville, VA 22901, USA

Cheng, Herbert S.
Director, Center for Engineering Tribology
Northwestern University
Evanston, IL 60201, USA

Dalman, David A.
Research Manager
Organic Specialty Laboratory
Dow Central Research
M.E. Pruitt Building
Midland, MI 48640, USA

Deresiewicz, Herbert
Department of Mechanical Engineering
Columbia University
New York, NY 10027-6699, USA

Eiss Jr., Norman S.
Department of Mechanical Engineering
Virginia Polytechnic Institute
and State University
Blacksburg, VA 24061, USA

Fein, Richard S.
Department of Mechanical Engineering
Columbia University
Consultant
35 Sheldon Drive
Poughkeepsie, NY 12603, USA

Fischer, Traugott E.
Department of Materials
Stevens Institute of Technology
Hoboken, NJ 07030, USA

Flom, Donald G.
Manager, Technology of Metal Cutting
Corporate Research and Development Center
General Electrical Company
P.O. Box 8
Schenectady, NY 12301, USA

Godet, Maurice
Institut National des Sciences Appliquees de Lyon
Batiment 113
20, avenue Albert Einstein
69621 Villeurbanne Cedex, FRANCE

Jahanmir, Said
Director, Tribology Program
Division of Mechanics, Structures
and Materials Engineering
National Science Foundation
Washington, D.C. 20550, USA

Kanakia, Mike
Southwest Research Institute
San Antonio, TX 78229, USA

Kennedy, Francis E.
Thayer School of Engineering
Dartmouth College
Hanover, NH 03755, USA

Kuhlmann-Wilsdorf, Doris
University Professor of Applied Science
University of Virginia
Charlottesville, VA 22901, USA

Landman, Uzi
Georgia Institute of Technology
Atlanta, GA 30332, USA

Larsen-Basse, Jorn
School of Mechanical Engineering
Georgia Institute of Technology
Atlanta, GA 30332, USA

Ling, Frederick F.
Rensselaer Polytechnic Institute and
Visiting Professor
Department of Mechanical Engineering
Columbia University
500 West 120th Street
New York, NY 10027-6699, USA

Nicholson, Richard
Principal Transmission Engineer
Rolls-Royce plc
P.O. Box 31
Derby DE2 8BJ, United Kingdom

Pan, Coda H.T.
Department of Mechanical Engineering
Columbia University
500 West 120th Street
New York, NY 10027-6699, USA

Peterson, Marshall B.
President, Wear Science Corporation
925 Mallard Circle
Arnold, MD 21012, USA

Ribarsky, Martin W.
Georgia Institute of Technology
Atlanta, GA 30332, USA

Somasundaran, Ponisseril
La Von Duddleson Krumb Professor of Mineral Engineering
Henry Krumb School of Mines
Columbia University
500 West 120th Street
New York, NY 10027-6699, USA

Tadjbakhsh, Iradj G.
Department of Civil Engineering
Rensselaer Polytechnic Institute
IPA - U.S. Army Research Office
Research Triangle Park, NC 27709-2211, USA

Tallian, Tibor E.
President, Tallian Consulting Corporation
36 Dunminning Road
Newton Square, PA 19073, USA

Tichy, John A.
Department of Mechanical Engineering,
Aerospace Engineering and Mechanics
Rensselaer Polytechnic Institute
Troy, NY 12180-3590, USA

Ting, Bond-Yen
Georgia Institute of Technology
Atlanta, GA 30332, USA

Vignoul, Gregory E.
Henry Krumb School of Mines
Columbia University
New York, NY 10027-6699, USA

Wedeven, Lavern D.
Principal Engineering Scientist
SKF Tribonetics
1100 First Avenue
King of Prussia, PA 19406, USA

Wert, James J.
Department of Mechanical
and Materials Engineering
Box 1621, Station B
Vanderbilt University
Nashville, TN 37235, USA

Winer, Ward O.
Regent's Professor
Georgia Institute of Technology
Atlanta, GA 30332, USA

Wu, Julian J.
Chief, Physics and Mathematics Branch
U.S. Army Research, Development
and Standardization Group
223 Old Marylebone Road
London NW1 5TH, United Kingdom

1. INVITED AREA LECTURES

1.1 FRICTION AND WEAR FROM THE MATERIALS SCIENCE VANTAGE POINT*

Gregory E. Vignoul

INTRODUCTION

Tribology, by its very nature, is a multidisciplinary subject which incorporates specialists in many different areas. However, the study of tribology has often polarized along set disciplinary lines, with each discipline rigidly adhering to its own terminology and research interests. Older explanations of friction and wear behavior have emphasized adhesive interactions of surface asperities. This approach has led to an emphasis on surface chemistry and on continuum mechanics, with much less attention given to the structures of the interacting materials themselves. Consequently, the purpose of this paper is to review more recent work which has studied the phenomena of friction and wear from a basic materials viewpoint. This approach focuses on the extensive plastic work in the near-surface region, transfer phenomena and sequences, mechanical mixing of contact layers and the fracture of these layers. It is then possible to explain the size of friction coefficients, energy dissipated, flake-like and non flake-like debris and microstructure effects.

FRICTION

This paper focuses largely on wear, but it is helpful to precede a review of wear models with a review of friction models. As with many problems in the physical sciences, friction can be approached by using forces or by using energy.

Perhaps the best known model for friction that utilizes the force approach has been postulated by Bowden and Tabor. In this model, they suggested that surface asperities adhere to form junctions and the friction force is directly related to the force needed to shear these junctions. The normal load L is supported by the total contact area A of the elastically and plastically deformed asperities. If τ_{jct} is the shear strength of the adhesive junctions, then the frictional force F is given by $F = A\tau_{jct} = (L/P_o)\tau_{jct}$, where P_o is the yield pressure (related to, but not equal to the hardness). Thus, the coefficient of friction $\mu = \tau_{jct}/P_o$. Estimates of the friction coefficient from this expression have been reported as $\mu = 0.2 - 0.3$.

Bowden and Tabor also included another contribution to friction, μ_p, due to the plowing of a metal by hard asperities. This term is calculated for different shapes of asperities by setting the friction force equal to the force needed to

---

*A summary of a lecture presented at the Workshop by Professor David A. Rigney of The Ohio State University.

displace material during the formation of grooves. Thus, the total friction coefficient would consist of two terms, an adhesion term μ_a and a plowing term μ_p, or $\mu = \mu_a + \mu_p$. Both terms would involve deformation. For smooth sliding, μ_a would dominate and for abrasion μ_p would be more important.

The principal results of the Bowden and Tabor model are reasonable; however, the model is too simple to be very useful. It is based on the assumption of homogeneous and isotropic materials, it uses a limited number of variables, and the parameters involved are somewhat ambiguous. Furthermore, basic materials characteristics, with the exception of a yield stress or a hardness parameter, tend to be neglected in most available models of friction. In particular, microstructural information is generally omitted. Yet, details of microstructure profoundly influence the nature of materials behavior, and they should be included in a realistic friction model.

Except for the lowest loads, some plastic deformation occurs near the surfaces of sliding materials, even under lubricated conditions. In fact, a principal mechanism for the dissipation of frictional work is through plastic deformation. Therefore, the ease of plastic deformation and the associated parameters such as those of the microstructure should be important considerations in setting up a friction model.

In an attempt to incorporate these ideas into an effective frictional model, P. Heilman and D. A. Rigney have presented an energy based model of friction. As mentioned before, this model focuses on the mechanisms by which energy can be dissipated by sliding. Specifically, it is assumed that for many materials the frictional work is approximately equal to the plastic work done, or $W_{pl} = \int \tau(\gamma)\Delta\gamma dV$, where $\tau(\gamma)$ is the shear stress, γ is the shear strain, and V is the deformed volume near the surface. This work includes energy stored as well as energy dissipated; the latter will dominate even for small plastic strains, and it accounts for essentially all of the plastic work at high strains. This approach seems particularly promising for metals, but it is probably also useful for some polymers and even for ceramics, which can deform plastically under compressive loading. The principal exception would be elastomers, for which hysteresis losses are mainly responsible for friction.

With this approach, energy terms are additive for each region and each material that is deforming; therefore contributions to the friction coefficient are also additive. Thus, one can readily treat all of the following cases: one material deforming (hard on soft), both materials deforming (similar hardness), coated materials, layers of reaction products, lamellar microstructures, etc.

To evaluate the work integral(s), one needs expressions for $\tau(\gamma)$, γ, and V. Both $\tau(\gamma)$ and γ will be functions of position. Also, $\tau(\gamma)$ is related to γ through a stress-strain relation. Since high strains are typically involved, a shear stress-shear strain relation similar to that of Voce is used in this model.

Specifically, $\tau = \tau_{max}(1-\exp(-c\gamma))^{1/2}$. This expression saturates at τ_{max}, has simple parabolic work-hardening behavior at low strains, and is consistent with simple dislocation arguments. Of course, for some materials, the stress-strain relation articulated above will not be a good representation for the stress-strain curve. For such cases, this model could still be used but different stress-strain relations, tailored to the particular interactions, would have to be used. Alternatively, a numerical representation could be used for complex stress-strain relations. It is important to realize that the use of a stress-strain relation is a convenient way to bring materials structure and properties into the friction model because the stress-strain relation reflects implicitly many basic characteristics of a material, e.g. the grain size and hardness.

While this model is not intended to give precise numerical results for the friction coefficient but rather to determine the dependence of the friction coefficient on familiar mechanical parameters as well as on microstructural features of a material, the model has been tested against experiment. This model was used to predict the friction coefficient of Cu sliding against a hard counterface. The friction coefficient obtained from the model was reported to be μ = 0.2(6). This value, when compared with that obtained experimentally by sliding OFHC Cu on 440C steel (μ = 0.3(0)-0.5(8)), gives a reasonable prediction of the friction coefficient, although an obvious disparity between the value predicted and the value obtained experimentally does indeed exist. It has been suggested that a possible source of this disparity lies in the transfer of material from the Cu to the steel (or vice-versa) which the simple friction model does not treat. However, the model can be extended to treated coated systems and even multiple layers.

WEAR

Wear is a complex subject. Even a brief survey of recent wear literature reveals a confusing variety of wear phenomena and proposed wear mechanisms. Indeed, it is common practice to consider wear in terms of a broad spectrum that ranges from abrasive wear to sliding wear (often referred to as adhesive wear). While the study of wear should ultimately provide insight into the many shades of this wear spectrum, it is sometimes sensible to select a manageable segment of the field for concentrated effort. Consequently, the following discussion will focus on the phenomena of sliding wear.

Given the restriction to sliding wear, the subject is still very broad, and it is desirable to look for trends and correlations which could be developed into useful wear relations. Unfortunately, this field is not yet at the stage where reliable wear equations or "laws" have been developed. Several simple wear models yield wear equations of the Archard type, but they are inherently limited by the fact that they rely on a single materials property (hardness), which is not sufficient. Also, it is clear from recent experimental work on a number of sliding systems that sometimes wear is not proportional to load and may not even be proportional to sliding distance.

At this stage, an approach which has much promise is one that focuses on the nature of wear debris. Most work in the past has emphasized the shape of the debris and the question of metallic or non-metallic debris. While these are important considerations, they must be supplemented by chemical and structural information. Furthermore, it is equally important to characterize geometrically, chemically and structurally the surface and subsurface of each material in the system.

With these observations and goals in mind, a series of sliding tests utilizing a wide variety of materials were performed. An experimental summary is presented below.

Tests and Conditions	Materials
- Block/ring, pin/disk	Cu, Cu-X X = Zn, Al, Si, Ge
- Room temperature	Cu, Cu-Ni, Ni
- Low speeds	γ - stainless steels
- Unlubricated	Dual phase steels
- Tests performed in vacuum	Cu-Be
or in Argon atmosphere	Cu-Cu_2O, Cu-Al_2O_3
	Cu, Ni, Fe, Mo
	Al, Al-X
	Coatings

A discussion of these studies will be presented in two sub-sections: Wear debris and Wear processes in sliding systems.

Wear debris: The generation of wear debris generally accompanies the wear of any sliding system. While a wide range of shapes and sizes of wear debris particles have been reported, a rough division of wear debris into two categories, flake and non-flake, can be made. Flake-type debris refers to wear debris that has a relatively uniform thickness whereas non-flake debris includes such debris types as plates, ribbons, cylinders, spheres, cutting chips, irregular chunks and loose clusters. The different types of wear debris observed in a system offer clues to the wear mechanisms involved, and they allow monitoring of the tribological health of the system by analysis of debris samples. It should be noted that both types of wear debris, flake and non-flake, were observed in this study.

Flake-type debris has received much attention in the literature. Clear micrographs are available showing this type of debris particle in the process of delaminating from a wear surface. The simplest explanation for this delamination is that subsurface cracks propagate parallel to the interface during sliding, and this optical observation has led to attempts to explain and predict delamination wear by applying deformation and fracture concepts to the base material. However, the evidence for direct delamination of base material to form flakes is very meager and unconvincing. This problem can be traced directly to the overreliance on the usual imaging mode of the scanning electron microscope. What is needed is careful characterization of the debris and of the near-surface material of the sliding components by an array of complimentary techniques such as energy-dispersive X-ray spectroscopy (EDS), X-ray diffraction, Auger electron spectroscopy, transmission electron microscopy (TEM) and others.

Indeed, application of these techniques in the sliding studies under discussion has provided strong evidence that the material which delaminates is not in fact the same, either structurally or chemically, as the base material. Instead, it is a very fine-grained material which may be derived from both parts of the materials system and may include components from the environment as well. This special layer of material is produced by a sequence which will be described in the section to follow, <u>Wear processes in sliding systems</u>.

Experimentally, the characterization techniques mentioned above showed that a common feature of the surface layer and the debris was a very fine grain size, ranging from about 3 nanometers to several tens of nanometers. For an OFHC copper block sliding on a 440C stainless steel ring, the small grains were mainly copper with a small proportion of α-Fe grains. For an age-hardened Cu-Be block sliding on an M2 tool steel ring, the small grains are again mainly copper, but with even smaller grains of beryllium oxide or copper oxide randomly distributed.

So far, this section has emphasized the structure of flake debris and of the layer from which it comes. However, if the hardness h of the processed layer relative to the hardness H of the adjacent base material is taken into account, it is then possible to predict when flake debris will be generated. If Tabor's discussion of the Mohs hardness scale is applied to this situation, it is expected that, for $h/H \leq 0.8$, the layer is not sufficiently hard to press into the base material and, therefore, it remains as elevated patches or plateaus on the surface. These are clearly visible in cross-sectional views of wear specimens. This form of layer material is associated with a rough and irregular sliding surface, and this in turn is associated with a "noisy" friction trace, i.e., the friction coefficient varies frequently and over a wide range of values. The mechanism by which the plateaus and patches yield flake debris is not yet certain, but the simple idea of a critical layer thickness is particularly promising.

Now consider non-flake debris. This debris typically has a pronounced lamellar structure and, in several ways, this type of debris is complementary to the flake-type debris already discussed. The non-flake debris is also derived from a very fine-grained layer, but one which is sufficiently hard, relative to the underlying material, to press into the substrate, i.e., $h/H \geq 1.2$. This creates a kind of hardfacing which influences both friction and wear. Associated with this non-flake debris is a smooth sliding surface whose main features are long but shallow grooves and ridges. In this case, sliding is smooth and the friction trace has much less noise. From experiments with Cu-Be alloys having various heat treatments, it is clear that the difference between this situation and that described with respect to flake debris need not be with the processed layer itself, since the structures observed by TEM and the microhardness values were found to be similar for the two cases. Rather, the critical parameter seems to be the relative hardness, which can change with different thermomechanical treatments of the bulk material, including those produced by sliding itself.

Wear processes in sliding systems: Much can be learned about wear processes by using tests which produce a macroscopic wear scar and numerous debris particles. However, these phenomena are the result of many local events, and the nature of these events is difficult to determine unless the very early stages of sliding are investigated. It is well established that transfer of material between contacting surfaces occurs in sliding systems, especially in the absence of effective lubrication. Since transfer modifies the surfaces of the sliding components, the friction and wear behavior can be affected. Therefore, studies of transfer phenomena are important for a fundamental understanding of friction and wear as well as for practical applications in engineering systems.

The direction of material transfer may be classified in three ways: one way transfer (A→B or B→A), mutual transfer (A→B and B→A) or back transfer (A→B→A etc.). Evidence for all three has been claimed in the literature but, as yet, there is no reliable theory to predict the direction and relative amount of transfer for a given system. The simplest theory states that transfer should occur if the shear strength of the adhesive bond between two asperities is greater than that of the transferring material. Thus, one might expect the "softer" material to transfer to the "harder" material, but this rule is not always obeyed. One complicating factor is the test geometry. From the work of Cocks and Antler it is known that the dominant direction of transfer may depend on the geometry of the sliding system. For a pin-on-disk system, transfer is preferentially from the disk to the pin, unless the unworn flat is much harder than the unworn pin. Therefore, the studies discussed here have included tests with the sliding components reversed, i.e., material B on A as well as A on B.

The tests in this study were divided into three categories. For category 1, the disk rotated for about a quarter of a turn. This allowed observation of initial transfer fragments. For category 2 (intermediate tests), the disk rotated for 15 cycles; these tests provided information on the evolution of transfer fragments. Finally, the tests in category 3 allowed correlation of transfer phenomena with friction and wear behavior after prolonged sliding (2500 cycles). This strategy allowed study of the development of surface features with time.

The sliding tests show that discrete fragments of material, ranging in size from a few micrometers to about 30 micrometers, transfer from one material to the other. Indeed, scanning Auger electron analysis of transfer fragments at various depths from their surfaces, obtained from the short-term tests (category 1), show that the transfer fragments are composed of material from one sliding component only, i.e., they are not yet blended with the material from their new hosts. It should be noted that these transfer events can occur after only a few micrometers of sliding. Further, mutual transfer (from pin to disk and from disk to pin) is a common feature of the systems studied.

Lamellar structures are often observed in the freshly transferred material. By using a stereo technique and parallax measurement, one can measure the average thickness of the individual transfer fragment layers. This technique, when applied to a transferred lamellar copper particle on an iron disk surface, gives a layer thickness of about 0.3 μm. This is the same as the average thickness of the subgrains observed adjacent to the surface in cross-sections of a copper block after prolonged sliding. Thus, at least for copper, there is a definite correlation between an important feature of transfer material and the substructure created by large plastic strains in the base material. In fact, it seems reasonable that the substructure produced by extensive plastic deformation near the wear surface provides the heterogeneity needed for shear instability, which leads directly to the formation of transfer fragments.

When transfer first occurs, the transfer material often looks much like a stack of coins. Soon after, or perhaps during, a transfer event, the lamellae are sheared in the sliding direction. With continued sliding, the transfer material is smeared, flattened and fractured, and it begins to blend with pieces of the new substrate material. This blending sequence is qualitatively similar to the early stages of the commercial process of mechanical alloying. In this process, various metal powders are placed in a high energy ball mill. During milling, the powder particles are deformed, fractured and intimately mixed by repeated compression and shear.

Examination of the wear surfaces generated in the intermediate tests (15 cycles) using SEM images and wavelength-dispersive spectroscopy (WDS) clearly show the lamellar features of recently transferred material. However, this examination also shows that 15 cycles are enough to have created a widespread system of mechanically mixed transfer patches. After prolonged sliding (2500 cycles), transfer patches develop into a mechanically mixed layer from which typical wear debris is generated. It is important to note that at long sliding times, such as those typically used to measure wear, various stages of the transfer and mixing processes described here are occurring simultaneously at different locations in the interface region.

As mentioned previously, the grain or subgrain sizes which develop in the transfer material of sliding systems can be quite small (less than about 10 nm). These sizes are much smaller than those of the microstructures which have been produced in bulk materials (about 1 μm). However, ultrafine grain sizes similar to those found in the sliding studies described in this paper have been observed in a variety of situations, although in most instances these ultrafine grain sizes occurred in two-phase systems. Thus, it may be that ultrafine grain structures need a second phase to stabilize them. The source of the second phase could be the base material itself, the counterface material, a coating or components in the environment. This requirement of a second phase seems reasonable, particularly given the microstructure obtained in self-mated systems such as copper sliding on copper in a clean environment. While this system produces a fine microstructure, the grain size is much larger than that observed in systems where two phases are distributed in the transfer material.

Friction data for the prolonged sliding tests were also collected. There seem to be three main types of friction behavior for this particular set of tests: friction traces with large fluctuations, smooth friction behavior, and transitions between the two. Comparison of the various friction traces reveals a correlation of considerable practicle importance. If a pin-disk sliding combination (A-B) produces a rough wear topography with large fluctuations of friction and large wear loss, then the reversed case (B-A) has a relatively smooth topography and friction trace and low wear loss.

The wear sequence presented here may be summarized as follows. Local contacts in sliding systems cause large plastic strains in either or both solid components. The plastic deformation changes the near-surface microstructure in ways which make the material unstable to local shear. This in turn produces transfer of pieces of deformed material which are further deformed and mixed with counterface material and/or environmental components to produce ultrafine grained material. The very fine microstructure in this transfer material is stabilized by the mixing in (mechanical alloying) of a second phase. The relative hardness of the transfer material and the adjacent substrate material affects the surface topography, the smoothness of sliding and the nature of the wear debris. Loose debris is commonly derived from the transfer material. There is some evidence that similar processes are important in lubricated systems.

Finally, geometrical factors and adhesion play important roles in the systems studied. Geometrically, the preferred transfer direction is from the disk to the pin, and hence a larger wear loss from the disk. From adhesion considerations, the preferred transfer direction is from the cohesively weaker material to the cohesively stronger. If one combines these two effects, a useful guideline is apparent: to reduce transfer and wear (and to maintain relatively smooth sliding) one should choose the cohesively weaker material as the pin and the cohesively stronger material as the disk.

REFERENCES

This paper has been compiled from notes taken during D. A. Rigney's presentation at the ARO Workshop, 18 December 1986. Also, portions of this paper have been incorporated from the references listed below.

D. A. Rigney and P. G. Shewmon, "Wear and Erosion of Metals," in *Metallurgical Treatises*, J. K. Tien and J. F. Elliott, editors, The Metallurgical Society of the American Institute of Metallurgical Engineers, New York, 621-641 (1981).

P. Heilmann and D. A. Rigney, "An Energy-Based Model of Friction and Its Application to Coated Systems, *WEAR, 72*, 195-217 (1981).

J. Don, T. C. Sun and D. A. Rigney, "Friction and Wear of Cu-Be and Dispersion-Hardened Copper Systems," *WEAR, 91*, 191-199 (1983).

Z. Y. Yang, M. G. S. Naylor and D. A. Rigney, "Sliding Wear of 304 and 310 Stainless Steels, *WEAR, 105*, 73-86 (1985).

L. H. Chen and D. A. Rigney, "Transfer During Unlubricated Sliding Wear of Selected Metal Systems," *WEAR, 105*, 47-61 (1985).

D. A. Rigney, L. H. Chen, M. G. S. Naylor and A. R. Rosenfield, "Wear Processes in Sliding Systems," *WEAR, 100*, 195-219 (1984).

APPENDIX. COMMENTS FROM D. A. RIGNEY, METALLURGICAL ENGINEERING, THE OHIO STATE UNIVERSITY.

During the planning stages of this Workshop, it was agreed that I would present one of the lectures, but someone else would prepare a written summary. I am delighted that Greg Vignoul was chosen for this task. His report is clear, comprehensive and well-written.

I would like to add just a few comments. The organizers of this Workshop encouraged us to bridge some of the differences between different approaches to friction and wear. This requires that we examine our own contributions more critically, and that we listen more carefully to what others are saying. Those of us with materials backgrounds must avoid the "pretty picture syndrome." If we present too much detail without adequate analysis, we create confusion instead of understanding. On the other hand, some of this detail is necessary if we are to develop more realistic models of tribological processes. We should not be content with traditional models if new data conflict with the assumptions of those models. For example, we should keep in mind recently identified processes as we try to develop improved wear equations. This is a challenging task, but we should make the effort to develop equations which are both useful and physically sound. Perhaps the concept of a rate-controlling step will help us to progress in that direction.

1.2 MODELING OF FRICTION AND WEAR PHENOMENA

Maurice Godet

I. INTRODUCTION

Modeling friction and wear is a challenge for the future, yet design engineers require that information now. Such information must be made available because of the considerable impact it will have on today's technology. The problem is to decide whether it will be obtained with the help of Tribologists or in spite of them. If Tribologists, concerned with dry friction problems and wear, acknowledge this need, they can still play a part in introducing their understanding in the modeling [1]. If not computer codes will continue, as they do now, to offer formulae chosen without discrimination, but which, because of their very existence, will be used in industry at large setting a trend which will take years to reverse.

The situation is different in other areas of Tribology and particularly in low pressure lubrication where efficient modeling has served to accurately describe the static and the dynamic characteristics of bearings [2-4] of various designs and allowed the bearing to be looked upon as an integral part of the machine structure [5].

Why is friction and wear modeling required by design engineers? First and foremost because "dry" contacts are as common as "wet" contacts and also govern the running qualities of machines. Let us look first at friction then at wear:

Friction: Design engineers need to understand how contacts, whether static or dynamic, modify the dynamic response of a structure. It is well known that flanges, pins, bolts, and even welds, introduce damping in mechanical assemblies which significantly alter their response. Unfortunately, a lot of effort is needed before satisfactory contact models are formulated and representative contact elements defined. And it will take time before these representative elements are introduced in classical finite element computer codes to predict assembly dynamic behaviour. Friction modeling, which conditions tangential compliance, appears to present the greatest obstacles to progress in this field.

Wear: Design engineers also need to understand how reliability, or design life, depends on wear. We cannot be satisfied today with the "change it when it is worn" attitude because other fields have made immense efforts to furnish system engineers or quality control specialists with data that can be understood and processed by reliability studies. Here again Tribologists will be left out of the picture and their know-how bypassed if wear is not expressed in terms palatable to that new race of men.

II. FRICTION AND WEAR MODELS

II.1 Common models

Friction: Normal compliance is dealt with, very satisfactorily, using techniques initiated by Kalker [6] and developed by others [7]. Tangential compliance [6] requires information on the contact shear stress or displacement distribution at the surface. Assumptions which cannot be verified must thus be made, and obviously, results are only as good as the assumptions. If in some solutions, no slip is permitted between the rubbing surfaces (adhesive condition), it is assumed in most cases that:

- the ratio "f" between local tangential and normal stresses, or local coefficient fo friction, is constant within the contact;
- "f" is a material property, which does not vary with time;
- slip only occurs when the tangential stress is larger than the product of "f" and the normal stress. This introduces a slip "threshold" and defines slip and adhesion areas in the same contact [9,10].

Wear: Formulae from early theories [11] are currently used in computer codes to calculate wear. These formulae suggest that the wear rate expressed in volume or weight lost per unit distance is proportional to the load divided by the material hardness. Here also, the constant of proportionality "k" is taken to be a material property.

II.2 Model appraisal

These elementary friction and wear models are much too "rustic" and are quite rightly subject to harsh criticism.

Friction: It is indeed easy to show that in many applications, the coefficient of friction varies significantly with load and speed, and that variations of the order of 4 to 1 [12-14] are common, particularly with polymers (Fig. 1 and Table 1). Further, satisfactory explanations based on thermal effects and variations in real area of contact have been advanced, but are difficult to express in terms which can be used in codes which use "rustic" friction descriptions.

Wear: Wear coefficients vary significantly and variations of the order of 10 to 1, which defeats their purpose, are not unusual (Table 1). Here also, qualitative explanations have been given but are not adaptable to computer "logic." The situation seems to have reached an impasse, because:

- there is, to my knowledge, little contact and no common language between tribologists in charge of wear and structural engineers in charge of computer code development;
- wear specialists are convinced, and sometimes with good reason, that their science cannot be modelled;
- structural engineers are not convinced that their models are not representative. They further observe that threshold concepts introduce unusual mathematical difficulties, [15,16] and feel that they have paid a heavy and

sufficient tribute to tribology. Little will happen on that front until something important and new turns up.

Clearly, the ball is in our camp, and mechanical engineers interested in wear could usually initiate a dialogue with structural engineers. The difficulty is to measure clearly the distance that separates us from a complete formulation of the problem and then to see what steps to take to bridge that gap. Let us say right from the start that the greatest step which can be taken today is to acknowledge the need for modeling and to keep this in mind in all of our research.

III. SOME BASIC OBSERVATIONS

As noted earlier, curves of variation of friction and wear vs. time, temperature, load, speed, geometry, environment, etc. and for different material combinations, are plentiful and at first sight should suffice. But, their very variations discourage most attempts at modeling. Clearly, a better understanding of the problem is necessary and reasons for these variations must be found. Let us start by following, stroke by stroke, a friction and wear test and try to identify as we progress, the factors that should be considered in model building. We have chosen a short stroke cycling test for three reasons:

1) it is representative of the type of situation found in the static assemblies discussed earlier, where modeling is most needed;
2) it is a low energy test and as such should not be perturbed by significant thermal excursions;
3) it lends itself to visualisation which has proved to be one of the most promising techniques in both lubrication and wear.

Fig. 2 [17] shows some of the pictures taken during a test in which glass was rubbed against steel. Observations were made every 10 runs. Fig. 2 shows that damage starts at the very first strokes and continues as time goes on. Observations taken later indicate that the steel and glass specimens are separated by a bed of debris and that the situation moved from a two-body (glass against steel) to a three-body (glass, debris bed or third body and steel) contact. In most instances, the debris bed acts as a protective coating [18]. Similar behaviours were obtained with other materials including purely metallic combinations, metals against polymers, ceramics, chalk, etc. [19-23] under both cyclic and continuous running. The early passage from a two- to a three-body contact appears to be quite general. The results which are presented below were obtained with three materials, two identical steel alloys (materials I and II) but quenched and tempered from different temperatures (TT) and a stainless steel (material III). Their composition in % is listed below:

MATERIALS

Material	C	Si	Mn	S	P	Ni	Cr	Mo	TT
I	.378	.33	.38	<.003	.011	3.77	1.65	.29	600°C
II	.378	.33	.38	<.003	.011	3.77	1.65	.29	200°C
III	<.03					12.5	17	2.75	

Friction: Fig. 3 [17] shows the change in coefficient of friction vs. both time (or number of tests) and displacement (or amplitude) during a characteristic cyclic wear test. The data presented, known as a "friction log," because of its shape, corresponds to a test run with alloyed steel specimens. The friction log is divided in four parts:

1) elimination of the "natural" screens that pollute the specimen surfaces;
2) increase in specimen (or first body) interaction accompanied by the corresponding first body structural changes (strain hardening, for instance, for metals);
3) debris formation and gradual passage from a two- to a three-body contact;
4) three-body contact characterized by a continuous formation and ejection of debris. Steady state conditions prevail [24].

The shape of the log depends, therefore, on [17]:

1) Surface preparation: in Fig. 4a, the steel specimens (material I) were cleaned first with acetone, then with alcohol, while in Figure 4b, they were cleaned with alcohol alone.
2) Materials: Fig. 5 shows the effect of tempering temperature (materials I and II) and composition (material III).
3) Frequency: Figs. 6a, b and c illustrate the effect of frequency.

Figs. 3 to 6 suggest that:

- Phase I is controlled by surface physics.
- Material properties, and particularly ductility, govern Phases 2 and 3.
- Machine dynamics control particle ejection and, as such, first-body protection.

Note that in all cases, steady-state conditions are reached after roughly 1000 cycles.

Wear: Some of these events were filmed by Y. Berthier with the help of the Service du Film de la Recherche Scientifique (S.F.R.S. Paris) and a run very similar to the one presented in Fig. 3, except that glass was rubbed against steel, was monitored. The film is taken through the glass specimen. Direct evidence of firstbody separation by a debris bed and consequently of debris load-carrying capacity was obtained. The process of debris formation and ejection is followed in great detail, and suggests that wear can be approached as a source and sink problem

[24]. Further, the debris can be replaced by a solid lubricant (a carbon fluoride was used) without causing much change in the operation. Normal wear started up again as soon as the solid lubricant was eliminated from the contact. Debris or solid lubricant action were difficult to tell apart.

It is obvious that modeling the friction logs presented in Figs. 3 to 6 is unrealistic today as the log shape is controlled by numerous parameters, many of which are only identified and not quantified. It seems, however, worthwhile as a first attempt to try to understand what happens:

- during the initial relative displacements between two solids in contact (for the first force displacement loop of Fig. 3, for instance);
- when steady-state conditions are reached, i.e. when the situation is somewhat better defined.

N.B. As this paper focuses on modeling of friction and wear, rather than on damage analysis, no attempt at explaining the physical or metallurgical aspects of the different forms of damage observed is made. It must, however, be clearly understood that even though steady-state conditions are encountered in runs as different as those described in Figs. 3 and 5c, the damage suffered by the specimen's bulk or first bodies is quite different and can lead to significantly different fatigue lives.

IV. MODELING FRICTION

Let us first recall that:

- it is necessary to understand the physics of friction resistance before attempting to model that resistance;
- friction force is the integral of the contact shear stress distribution.

IV.1 Initial displacements

The variation of contact stress distribution of a birefringent rectangular slab loaded against a flat during a tangential load cycle was monitored using photoelastic techniques [25-28]. The tangential load was increased until full slip was encountered and decreased gradually until it reached zero. This experiment was performed to understand what happens to the stress field at contact just before and during initial displacements, i.e. during the first cycle of Figs. 3 to 6. Specifically, information on the factors which govern boundary conditions in a two-body contact was sought.

R. Progri and B. Villechaise filmed the event with the help of the S.F.R.S.. The loading pattern is described in Fig. 7. The normal load N is first applied, then the tangential load T is imposed gradually. As soon as global relative slip is observed between the two bodies, the tangential load is relaxed. Three different behaviours were noted during the loading phase:

1) Full adhesion between the two loaded surfaces was observed during the entire normal load application and during the beginning of the tangential

loading.

2) Partial slip was noted as the tangential load was increased and was characterised by the passage of waves [29] which swept part of the contact width.

3) Full slip occurred at higher tangential loads, as the perturbation swept the entire contact width causing relative displacement between the two specimens.

Phase 1, which is reversible, could be modeled using classical elasticity theory with adhesive (no slip at the boundary) conditions. Theoretical and photoelastic stress fields were within experimental error.

The same approach used in Phase 2, after the passage of waves, did not yield such good results. However, a satisfactory accounting of the experimental stress field was obtained by assuming that each of the perturbations corresponds to the passage of an interfacial crack in Mode II, suggesting that initial slip could be explained with fracture mechanics concepts.

It is no longer possible theoretically to adjust the parameters of the crack models during full slip, or Phase 3, as not a trace of the phenomena is left because the entire contact is swept by the waves. However, crack intensity factors can be determined by analysing each frame of the film.

These conclusions are encouraging as they suggest that fracture mechanics might help model initial displacements. This would be particularly useful as none of the evidence produced in this study justifies the classical Coulomb friction law as the ratio of shear to normal stress varies across the width of the contact in all of the experiments. Great care, however, has to be exercised before extrapolating the results described here to other material combinations, but wave propagations have been observed with many other materials, which suggest that the phenomena reported here possess some generality

Note also that two different stress fields were measured for the same external normal load, before and after the tangential load was applied and removed. This occurred because the surface relative displacements do not follow the surface stresses, threshold effects complicate the problem. Results suggest that the entire stress tensor at the surface, and not only the stress vector, as suggested by the Coulomb boundary conditions, must be considered in a complete description of the contact (Fig. 8). Thus, the stress distribution at the surface, which is the expression of the boundary conditions is governed by the response of the entire slab to the external loads and not only by surface physics.

IV.2 Steady-state conditions

Fig. 3 shows that the shape of the friction log does not vary significantly beyond 500 strokes. The friction cycle is well established, the process, which is apparently easy to model, if averaged values are acceptable, has settled down to a steady-state condition. The situation is unfortunately not quite as simple because:

- the coefficient of friction varies easily by a factor of two from one metal combination to another and higher variations can be obtained with other material combinations;
- the coefficient of friction varies with contact geometry, speed and temperature;
- the number of cycles necessary to achieve steady-state conditions depend on the materials tested and particularly on their ductility.

Modeling, whose function is predictive, therefore requires the kind of information contained in Fig. 3 to have a chance to get close to reality. This is, of course, only possible if some generality can be given to that figure. The problem will be approached from two angles, one systematic, the other fundamental:

Systematic: Extrapolation in tribology [30] is known to be a very perilous exercise, and a lot of effort has gone into explaining why this is so. While this "campaign" was and still is absolutely necessary to inform design engineers of the dangers of material classification in tribology, it might be advisable today to identify a few representative types of contacts and determine the extent of the variation in friction coefficient. More specifically, let us start quantifying the variation, of friction forces under steady-state conditions, of a:

- plain structural steel quasi-static assembly
- composed of a flat on flat contact
- exposed to normal ambient conditions (rain excluded)

discussed earlier (III) with contact size, shape defect, load, frequency, and (small) relative displacement. Such a program could produce reasonably rapidly "interpolation charts" which would be of the highest use to designers, and would establish and formalise the much needed relations between tribologists and design engineers. The latter would be very receptive to this kind of information and approach as, in time, it could be made compatible with C.A.D. systems.

One must recognise that this attitude is scientifically dangerous, and strict controls are needed to ensure that the data produced will only be interpolated and not extrapolated.

At first sight, the systematic type of research suggested here might not appeal directly to the scientific community at large but:

- the quality of the information needed requires very elaborate equipment whose cost could only be justified by such a program;
- the research, conducted on this elaborate equipment, need not be limited to the determination of friction cycles as will be seen later.

Fundamental: The systematic research discussed above can produce rapidly global values of the coefficient of friction but it brings no insight on normal and tangential stress distributions which is expected of a complete three-body contact analysis. Such an insight will only come from:

- a thorough understanding of adhesive conditions between powder beds (or third bodies) and first body surfaces, which will condition boundary slip [31] and thus the boundary conditions introduced in models;
- a thorough understanding of the behaviour of more or less cohesive bead chains [32] or of more or less cohesive powder beds [31,33] under the very high strain rates imposed in contacts, i.e. an understanding of powder bed or third-body rheology (stress vs. strain and time) under contact conditions.

Let us therefore look at both boundary conditions and third-body rheology under steady-state conditions:

a) boundary conditions

Adhesion between powders and solids [34] is a difficult subject and elementary tests [24,31] show that:

- great differences are noted between powders of identical composition but of different grain size;
- adhesion can thus vary during a wear test as particles detached from first bodies are ground to smaller sizes.

Clearly, the definition of representative boundary conditions is certainly one of the great obstacles on the road to modeling steady-state conditions [35]. Close connections with surface physicists are needed to advance in this domain [36].

b) third-body rheology

Two series of tests which were run quite independently of each other and with different objectives give some indication on powder rheology:

1) In one of the series of tests, average normal and shear stresses vs. average shear rates were obtained for cohesionless carbon-steel beads [33] under light loads, reasonably high continuous speeds, and, (for wear tests), very thick films. The test apparatus consisted of two concentric aluminum disks mounted on a vertical rotating shaft. The upper disk is ribbed, the driven lower one channeled and the sheared powder is held in the channel. Friction torque and disk separation are monitored. Boundary slip is controlled. Results show that for constant fractional solid contents (apparent density), the stresses vary as the shear rate raised to a power close to two.

2) In the other series of tests, powder beds were shown to be able to call upon at least three mechanisms [22] to accommodate the velocity differences between first bodies:

- boundary slip
- rupture
- continuous velocity gradients across all or part of the film thickness.

Fig. 9 gives the rupture fronts and the velocity profiles observed by following, through X-ray monitoring, the displacements of small lead balls placed in a cohesive (when compared to the one discussed above) bed entrained under a fixed inclined plane. The test device is illustrated in Fig. 10. Here also, the film thickness is much larger than those found in friction and wear tests and it seemed necessary to examine what happens under conditions much closer to reality. Y. Berthier, again with the help of the S.F.R.S., filmed a dynamic contact in which the first bodies were separated by a more representative graphite film of a few tens of microns in thickness. The film clearly shows that the three velocity accommodation mechanisms are found in such a contact, and that even "turbulence" is observed as in lubrication [37,38].

V. MODELING WEAR

Most treatises [39] on wear concentrate on the mechanics of particle detachment from first bodies and relate wear rate to the rate of particle detachment. Particles are detached according to the following basic mechanisms [40,41]:

- adhesion
- abrasion
- fatigue

to which corrosion and possible other effects should be added. Elasticity [7,8], plasticity [42,43] and fracture mechanics studies [26,44] have shown how first-bodies respond to external loads, and thus indirectly explain particle detachment.

The third-body approach (TBA) presents things somewhat differently in suggesting that wear rate is governed by three apparently independent factors [20]:

- particle detachment
- third-body properties
- environmental dynamics.

The TBA notes that, once detached, particles are momentarily trapped (dwell times), for purely geometrical reasons (Fig. 11) and argues that a small particle (10^{-6}m) cannot be instantly eliminated from a comparatively wide contact (10^{-2}m), causing the situation to move gradually but rapidly from a two to a three-body contact as indicated in Figs. 2 and 3. Clearly, third-body dwell times, or its opposite, third-body flow, depend on contact shape, kinematics, contact dynamics, and particle bed, or third-body properties. A wide, quiet, low frequency, fretting contact which generates cohesive debris which adhere to the rubbing surfaces will present much higher dwell times than a narrow, trepidant, high speed, continuous contact which produces noncohesive debris that do not stick to the first bodies. In simpler terms, the first type of contact traps debris while the second eliminates them. If it is confirmed [42,18] that, contrary to what is usually believed, debris have load-carrying capacity [18] and that the presence of debris is often beneficial from a wear point of view, wear rate will decrease with trapping ability.

If, as suggested by the TBA, wear is governed by the three independent factors (particle detachment, third-body properties, environmental dynamics) mentioned above, theories based on particle detachment alone could only describe wear, when particle detachment dominates over the two other factors. Studies must therefore be undertaken to decide when one of these factors dominates the other. Luckily, elementary considerations on third-body dwell times lead us rapidly to first conclusions.

Short dwell times imply empty contacts, high third-body flow and thus no third-body protection, limited first-body interaction, and low wear.

This analysis leads us to compare the kinetics of two processes or two rates:

- the rate of third-body formation
- the rate of third-body elimination, or third-body flow.

The highest rate of third-body formation is obviously obtained for zero dwell times and the lowest, which can reach zero, for infinite dwell times. Dwell times, which depend on the rate of elimination, thus govern third-body formation. As in a steady-state process, the two rates are equal, dwell times control wear. Wear modeling therefore should be centered on third-body dwell times or third-body flow. The two terms, "dwell times" and "flow," were chosen hoping that the first will appeal to material scientists and the second to mechanical engineers. A high dwell time is obviously synonymous with low flow.

Two conclusions can be drawn from this analysis:

1) Wear is governed by contact dwell time. The parameters which control dwell times must be identified to model wear. More concretely, wear is controlled by particle ejection.
2) It is correct to say that first-body particles can be produced through adhesion, abrasion, fatigue, etc., but it is not correct to talk about wear by adhesion, abrasion, fatigue, etc., as that list is insufficient to describe the entire wear process.

VI. DWELL TIMES AND FLOW UNDER STEADY-STATE CONDITIONS

Identifying today third-body dwell time as the governing parameter which has to be considered in any attempt at modeling wear, is certainly very helpful as it focusses the attention on one aspect of the problem which encompasses surface physics, material science and mechanics, which we know play a part in this very complicated and elusive process. That advantage must not, however, hide the fact that it is a composite parameter, which has no intrinsic scientific value as, unlike flow, it cannot be measured. It might, however, encourage scientists of different fields to get together and attempt to break down that parameter into its many components and try to rank them in importance in any given well defined contact situation.

In this presentation, the idea that the rubbing specimens [24] will continue to wear unless separated by a discrete or continuous third-body does make sense. How much third-body is needed and how thick for given materials and mechanical conditions must it be is, however, unknown. If flow is impeded (i.e. if the rate of third-body elimination is decreased) wear will drop. Third-body elimination, or wear, thus depends as much on the geometric and dynamic environment as on the behaviour of the material itself.

Table II tries to identify the parameters which control "dwell time" or "flow" in terms of Mechanics and Materials Science. Flow is said to depend on:

- boundary conditions,
- rheology, which, for a given material, depends on pressure, temperature, and shear-rate.

Dwell times are said to depend on:

- surface science which brings to mind surface energy and adhesion concepts,
- powder behaviour expressed in terms of cohesion, specific surface, reactivity, adsorption, etc.

The two discourses are by no means antinomic; they present no intellectual or scientific hierarchy but are definitely nonconvergent even when put face to face. They each follow their own line of reasoning and will naturally go on doing so unless one of the partners decides that, due to unforeseen events, he cannot go on single handed anymore and "interpellates" the other. The one who "interpellates" today is the Mechanical Engineer, and the unforeseen event is the obligation to model Friction and Wear. He has to learn to go back and forth between the left and right columns of Table II and find out how, for instance, surface energy affects tangential slip between the powder bed and the specimen surface, or how adsorption, for instance, favors third-body rupture fronts over continuous deformation fields. He cannot be led astray by relations between specific surface and cohesion unless he is told how cohesion governs the mechanisms of velocity adaptation across the third-body. His leadership is imposed by the subject but is it not inevitable in other chapters of Tribology where he can be content with a more humble role.

VII. DISCUSSION AND CONCLUSION

The argument developed in this paper confirms that neither friction nor wear are sufficiently understood to offer a solid basis for modeling. It is, however, useful to raise the question if only to restate the necessity of having models, to take stock of the limitations of those in use today, and insure that some of the new work can be redirected with the idea of modeling in mind.

VII.1 Answers

Simulation: Some answers can be given to Design Engineers if simulation campaigns, in which all the factors listed in this study are taken into account, are set up. A good place to start would be to measure both the wear rates and the stiffness and damping characteristics of flat on flat assemblies of well identified materials subjected to controlled vibrations in a controlled atmosphere. This must be done in liaison with a design engineer:

- to ensure that the data produced can be integrated easily in computer codes;

- to develop a "feel" for the type of accuracy needed to give meaning to the design calculation and appreciate if that accuracy is compatible with the variations measured over time in a representative industrial assembly.

The campaign must also be run with the help of a material expert who can follow the change in third-body composition over time, and the effect of that change on the stiffness and damping characteristics of the simulated assembly.

Other solutions: The program suggested above attempts to characterise the solutions which are commonly used today. The problem could also be approached differently, by looking into new solutions which offer better controlled contact situations than those found with classical materials. It is not impossible to think in terms of inserts, coatings, or solid lubricants which could produce a "predictable" contact with fairly constant properties over time. Design engineers would welcome such suggestions. The program presented above could easily be extended to include such an approach.

VII.2 Fundamentals

Limits: Fundamental studies must also be continued but it seems wise to limit modeling attempts to the initial two-body or the steady-state three-body problems and leave the two- to three-body passage aside for the time being.

Fracture mechancis: The degree to which fracture mechanics can serve to describe initial displacements and thus initial stress fields must be identified if fracture mechanics are to be used in modeling.

Powder composition and rheology: Even under steady-state conditions, the same first-body metal combination will produce different third-bodies, with different properties under different loads, speeds, temperature, atmosphere, etc.. These changes in composition and the effect of these changes in composition on properties have to be identified, before the data produced can be organised and used in computer codes. For instance, powder mechanics studies under high pressure and high shear-rates and different atmospheres must be undertaken rapidly if only to be convinced of the extreme complexity of powder rheology in itself and of the great variations in behaviour that apparently minor changes in pressure or atmosphere can engender. Such work must necessarily be done in very close liaison with material scientists and high pressure rheology experts.

Wall friction: Wall friction control is needed. One must also decide whether adhesion between first and third-bodies is always desirable. A first guess would be yes when relative displacements are microscopic as high adhesion would favour trapping. But the situation could be quite different if the displacements are large as high adhesion would entrain third-bodies outside of the contact and provoke wear.

In conclusion, tribologists must keep modeling in mind in future studies, and set up cross disciplinary programs with material scientists (volume and surface experts), mechanical engineers (applied mechanics and design), rheologists, and modeling specialists. This is a tall order, but it can be tackled by a step by step approach and could prove to be vital for the credibility and survival of Tribology.

REFERENCES

1. "Frictional Damping at Preloaded Metallic Interfaces Under Tangential Loading," B. B. Aggarwal, R. A. Russell and W. R. D. Wilson, in *Friction and Traction*, D. Dowson, C. M. Taylor, M. Godet and D. Berthe, editors, Westbury House, Guildford, Great Britain (1981).
2. "Comparison of Dynamic Behaviour of Elastic Connecting Rod Bearings in Both Petrol and Diesel Engines," B. Fantino and J. Frêne, *Journal of Tribology, Transactions of the American Society of Mechanical Engineers*, *107*, 87-91 (1985).
3. "Rotor Bearing Dynamics," C. H. T. Pan, in *Tribology*, A. Z. Szeri, Hemisphere Publishing Corporation, McGraw-Hill Book Company, New York (1980).
4. "Stability and Unbalance Response of Large Turbines," N. Abdul-Wahed, D. Nicolas and M. T. Pascal, *Journal of Lubrication Technology, Transactions of the American Society of Mechanical Engineers*, *104*, 66-75 (January 1982).
5. "Dynamic Analysis of Engine Bearing Systems," W. A. Welsh and J. F. Booker, in *Tribology of Reciprocating Engines*, D. Dowson, C. M. Taylor, M. Godet and D. Berthe, editors, Butterworths, London (1983).
6. "A Minimum Principle for Frictionless Elastic Contact with Application to Non-Hertzian Half-Space Contact Problems," J. J. Kalker and Y. van Randen, *Journal of Engineering Mechanics*, *6*, 193-206 (1972).
7. "Influence of Surface Waviness and Roughness on the Normal Pressure Distribution in the Hertzian Contact," J. Seabra and D. Berthe, to be published in *Journal of Tribology, Transactions of the American Society of Mechanical Engineers* (1987).
8. *Contact Mechanics*, K. L. Johnson, Cambridge University Press, Cambridge, 202 (1985).
9. "Compliance of Elastic Bodies in Contact," R. D. Mindlin, *Journal of Applied Mechanics, Transactions of the American Society of Mechanical Engineers*, *16*, 259 (1949).
10. "Aspects of Contact Mechanics," J. J. Kalker, in *The Mechanics of the Contact Between Deformable Solids*, Delft University Press, Delft, The Netherlands (1975).
11. "Contact and Rubbing of Flat Surfaces," J. F. Archard, *Journal of Applied Physics*, *24*, 981-988 (1953).
12. "Basic Mechanisms of Friction and Wear of Polymers," J. K. Lancaster, in *Plastics and Polymers*, *41*, 297–306 (1973).
13. "Design of High Performance Dry Bearings," D. Play and M. Godet, *WEAR*, *41*, 25-44 (1977).
14. "Simulating Contact Conditions in Dry Bearings," D. Play, *Tribology International*, 295-301 (October 1978).
15. *Les inéquations en mécanique et en physique*, G. Duvaut and J. L. Lions, Dunod, Paris (1972).
16. "Stress Wave on the Contact Area of a Solid Sliding with Friction on an Obstacle," M. Raous, F. Taallah and B. Villechaise, *Euromech 209: Vibrations with Unilateral Constraints*, Como, June 5-7, 1986.
17. "Usure induite sous petits débattements," Ch. Colombié, Thèse de Doctorat, Ecole Centrale de Lyon, Lyon, September 1986.
18. "Fretting: Load Carrying Capacity of Wear Debris," Ch. Colombié, Y. Berthier, A. Floquet, L. Vincent and M. Godet, *Journal of Tribology, Transactions of the American Society of Mechanical Engineers*, *106*, 185-194 (1984).
19. "Self Protection of High Wear Materials," D. Play and M. Godet, *American Society of Lubrication Engineers Transactions*, *22*, 56-64 (1979).
20. "The Third-Body Approach: A Mechanical View of Wear," M. Godet, *WEAR*, *100*, 437-452 (1984).
21. "Mécanique de la Tribologie, applications aux céramiques," M. Godet and Y. Berthier, *L'Industrie Céramique*, No. 808, 9, 9/86.
22. "Effet du comportement du troisième corps sur l'usure des paliers secs en

mouvement alternatif," Y. Berthier, Thèse Docteur Ingénieur, I.N.S.A. de Lyon, 9 Juillet 1982.
23. "Wear Mechanisms in Oscillating Bearings," Y. Berthier and D. Play, *WEAR*, *75*, 369-387 (1982).
24. "First and Third Body Effects in Fretting - A Source and Sink Problem," Y. Berthier, Ch. Colombié, G. Lofficial, L. Vincent and M. Godet, in *Mechanisms and Surface Distress*, D. Dowson, C. M. Taylor, M. Godet and D. Berthe, editors, Butterworths, London (1986).
25. "Contribution à l'étude théorique et expérimentale des conditions aux limites d'un contact sec avec frottement," R. Progri, Thèse de Doctorat-ès-Sciences, Paris VI, 1984.
26. "Mécanique des Contacts: Elasticité et Rupture," B. Villechaise, Thèse de Doctorat-ès-Sciences, INSA-UCB, Lyon, December 1985.
27. "Boundary Conditions in a Two-Body Contact Formed by a Rectangular Polyurethane Slab Pressed Against an Araldite Plane," R. Progri, B. Villechaise and M. Godet, *Journal of Tribology*, *Transactions of the American Society of Mechanical Engineers*, *107*, 138-141 (1985).
28. "Fracture Mechanics and Initial Displacements," R. Progri, B. Villechaise and M. Godet, in *Mechanisms and Surface Distress*, D. Dowson, C. M. Taylor, M. Godet and D. Berthe, editors, Butterworths, London (1986)
29. "How Does Rubber Slide?" A. Schallamach, *WEAR*, *17*, 301-312 (1971).
30. "Extrapolation in Tribology," M. Godet, *WEAR*, *77*, 29-44 (1982).
31. Y. Berthier, Thèse Doctorat-ès-Sciences (in preparation).
32. "Stability of Chains of Permeable Spherical Beads in an Applied Magnetic Field," R. S. Paranjpe and H. G. Elrod, *Journal of Applied Physics*, *60*, 418-422 (1986).
33. "An Experimental Study of the Rapid Flow of Dry Cohesionless Metal Powders," K. Craig, R. H. Buckholz and G. Domoto, to be published in *Journal of Applied Mechanics*, *Transactions of the American Society of Mechanical Engineers*.
34. "Powder Wall Friction," S. Strijbas, *Powder Technology*, *18*, 209-214 (1977).
35. "An Attempt to Provide a Unified Treatment of Tribology Through Load Carrying Capacity, Transport, and Continuum Mechanics," M. Godet, D. Play and D. Berthe, *Journal of Lubrication Technology*, *Transactions of the American Society of Mechanical Engineers*, *102*, 153-164 (1980).
36. "Influence of the Metal Substrate on the Formation of a Graphite Transfer Film," M. Brendlé and J. Fatkin, in *Mechanisms and Surface Distress*, D. Dowson, C. M. Taylor, M. Godet and D. Berthe, editors, Butterworths, London (1986).
37. "Turbulence in High Speed Journal Bearings," D. F. Wilcock, *Transactions of the American Society of Mechanical Engineers*, *72*, 825-834 (1950).
38. "Flow Transition Criteria in Journal Bearings," J. Frene and M. Godet, *Journal of Lubrication Technology*, *Transactions of the American Society of Mechanical Engineers*, *96*, Series F, 135-140 (1974).
39. "Friction and Wear," I. V. Kragelsky, M. N. Dobychin and V. S. Lombalov, *Calculation Methods*, Pergamon Press, Oxford (1982).
40. "Visualization of Chalk Wear," D. Play and M. Godet, in *The Wear of Non-Metallic Materials*, D. Dowson, M. Godet and C. M. Taylor, editors, MEP, London (1977).
41. "An Overview of the Delamination Theory of Wear," N. P. Suh, *WEAR*, *44*, 1-16 (1977).
42. "Plasticity Analysis of Wear Phenomena in the Development of Abradable Materials," F. E. Kennedy, K. E. Hauck, L. P. Grotelueschen, in *Mechanisms and Surface Distress*, D. Dowson, C. M. Taylor, M. Godet and D. Berthe, editors, Butterworths, London (1986).
43. "Extrusion, a New Wear Mechanism," A. V. Olver, in *Mechanisms and Surface Distress*, D. Dowson, C. M. Taylor, M. Godet and D. Berthe, editors, Butterworths, London (1986).
44. "A Fracture Mechanics Correlation with Fretting Fatigue Experiments," J. J. O'Connor and D. A. Fields, in *Mechanisms and Surface Distress*, D. Dowson, C. M. Taylor, M. Godet and D. Berthe, editors, Butterworths, London (1986).

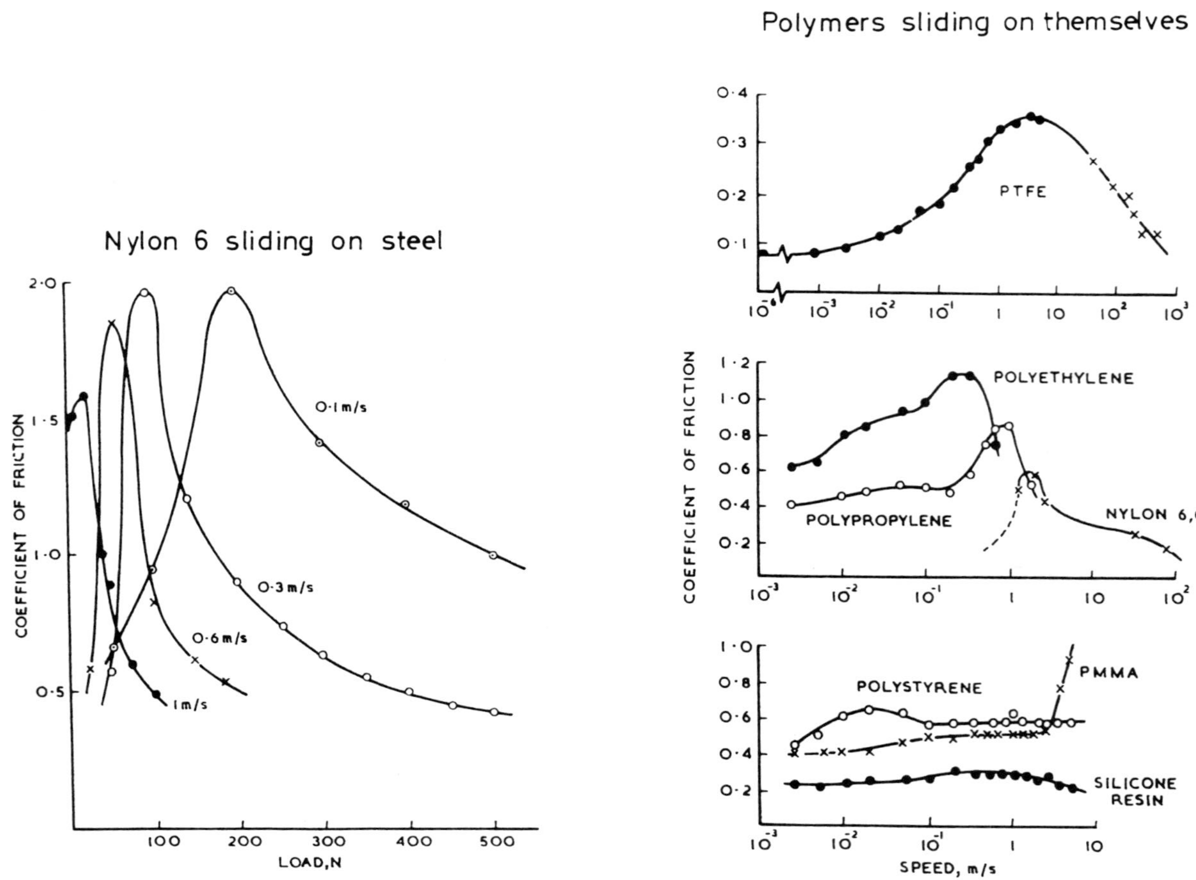

Fig. 1. **Variation of Coefficient of Friction with Load and Speed. (Works quoted in Reference 12, p.299.)**

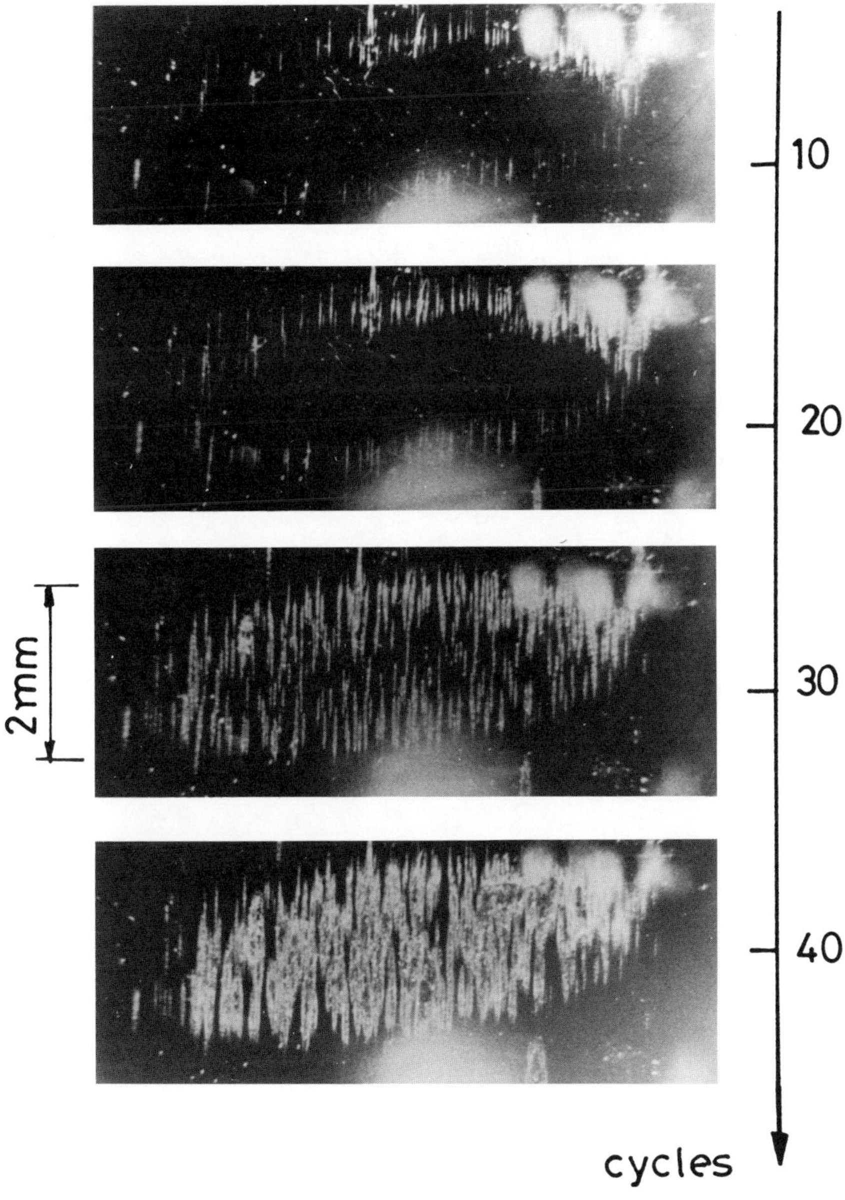
10
20
30
40
cycles
2mm

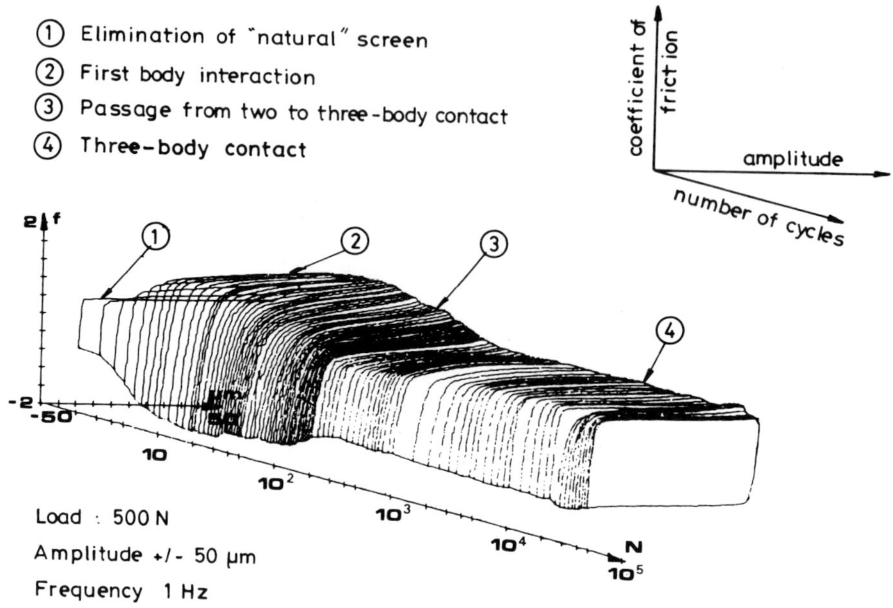

Fig. 3. Friction Log Obtained with Material 1 (Reference 17, p.22).

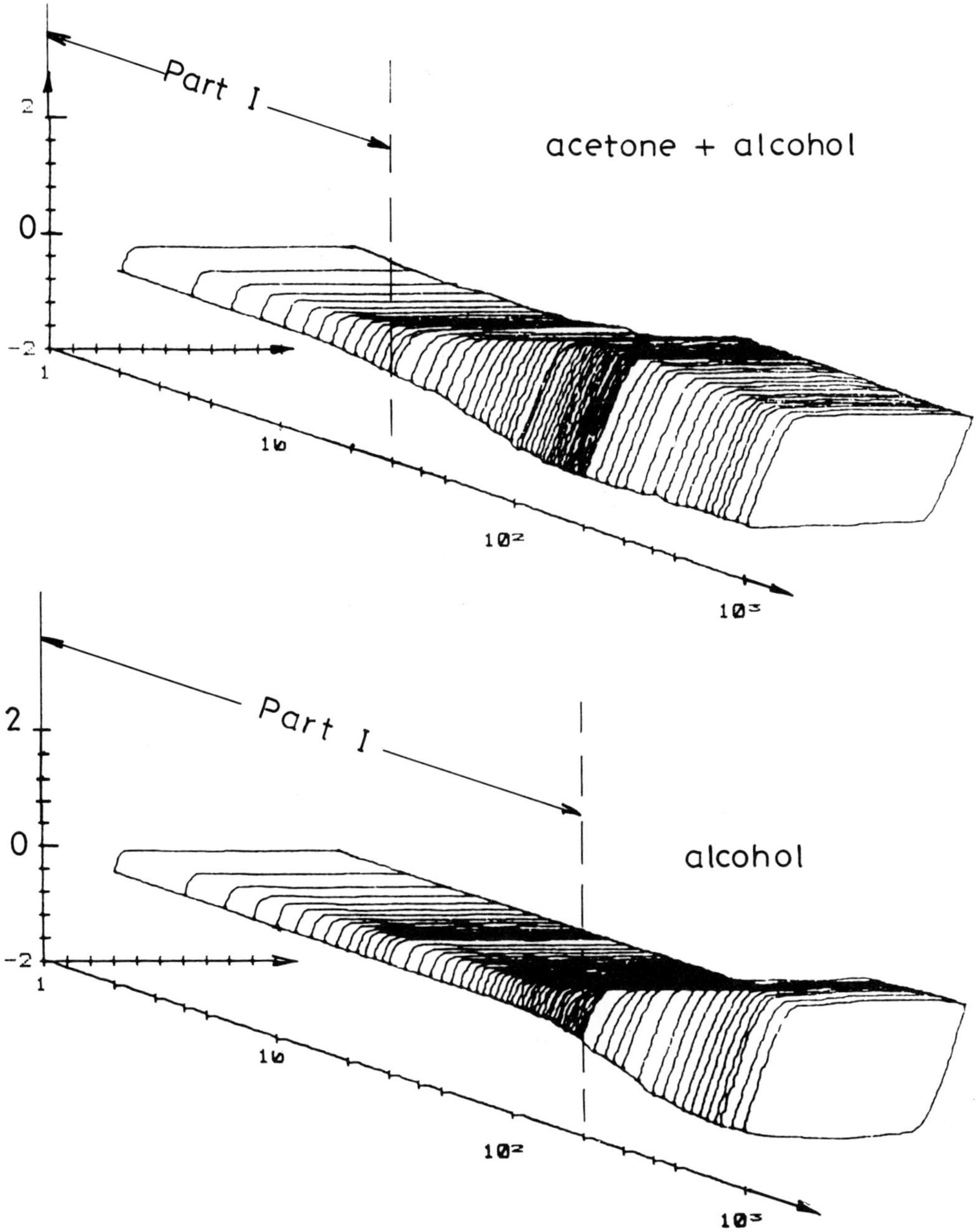

Fig. 4. Effect of Specimen Preparation (or Cleaning) on Friction Log (Reference 17, p.41).

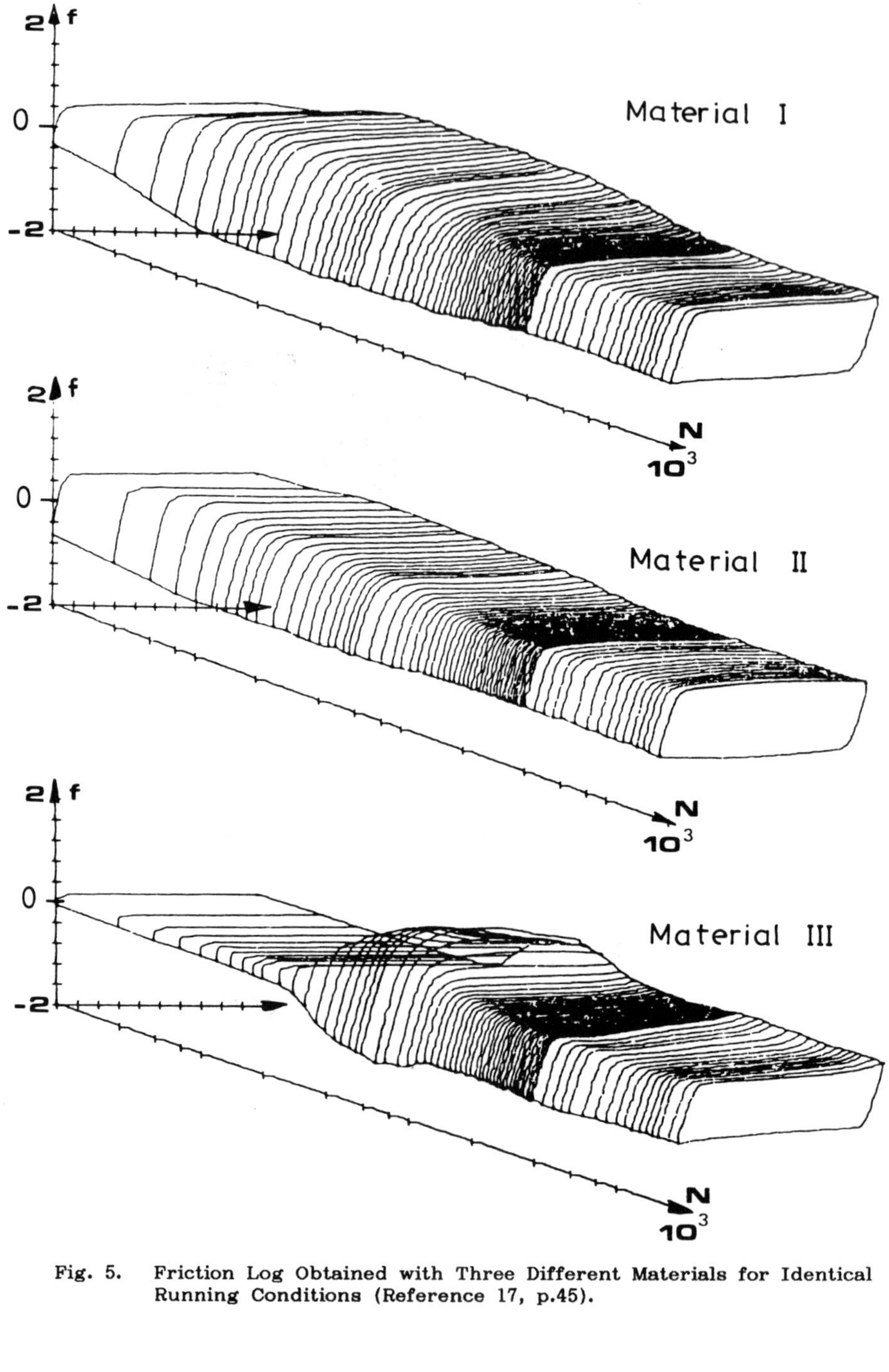

Fig. 5. Friction Log Obtained with Three Different Materials for Identical Running Conditions (Reference 17, p.45).

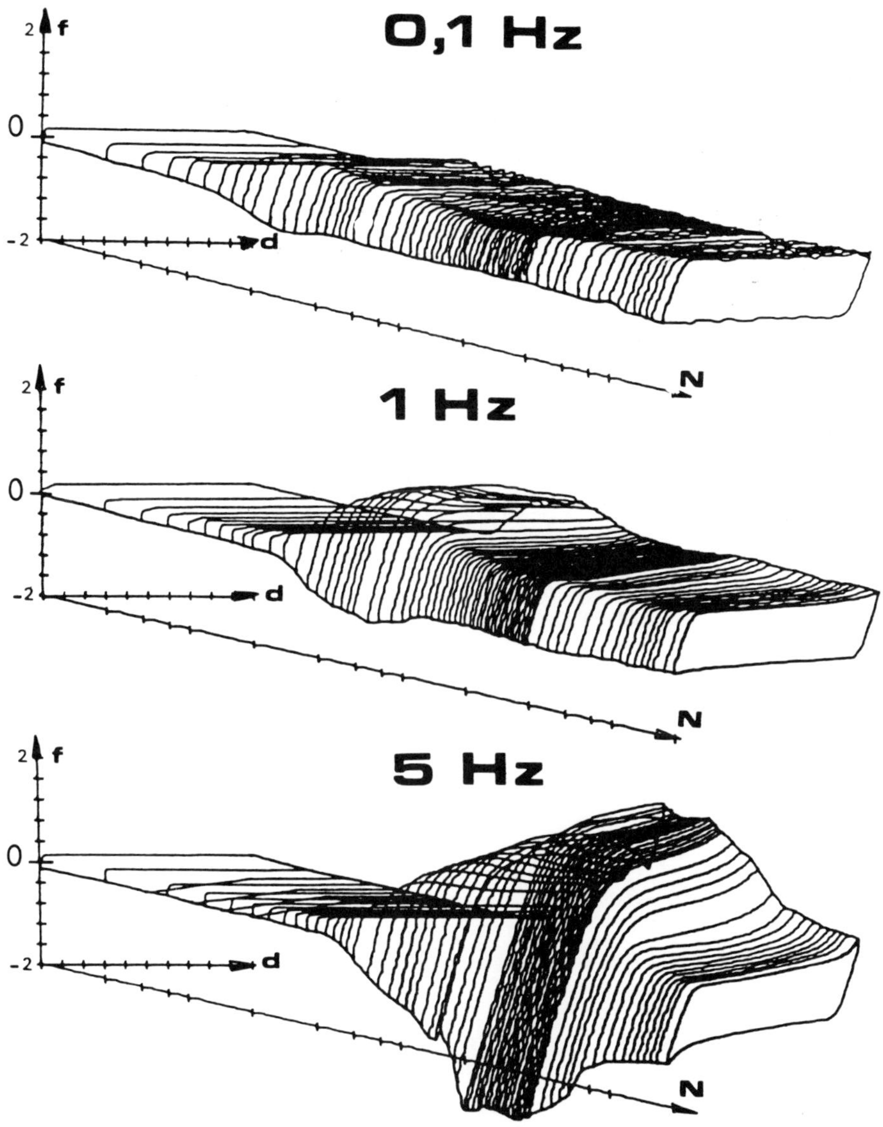

Fig. 6. Effect of Frequency on the Friction Log (Reference 17, p.60) for Material III.

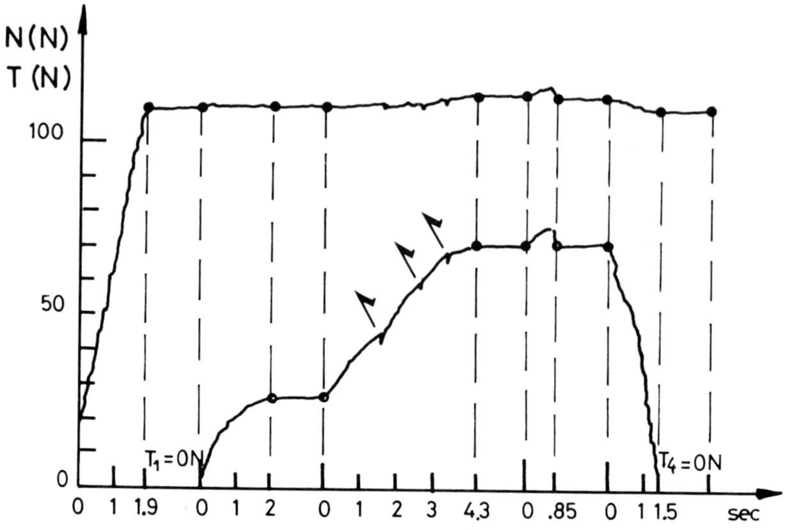

Fig. 7. **Normal and Tangential Load Cycle (Reference 28). (Flags correspond to the passage of perturbations.)**

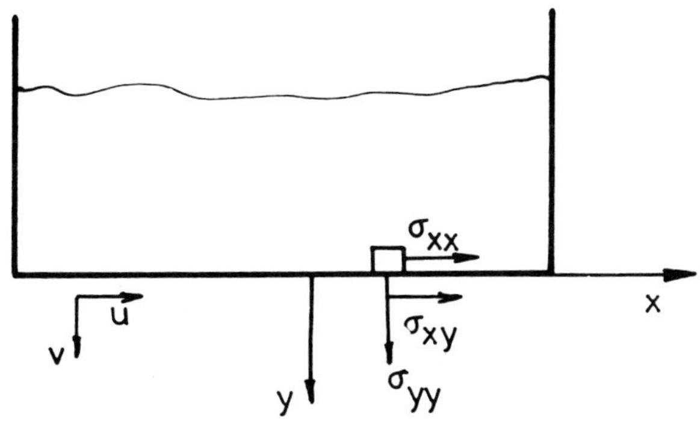

Fig. 8. σ_{xy} and σ_{yy} Define the Stress VECTOR,
σ_{xy}, σ_{yy} and σ_{xx} Define the Stress TENSOR.

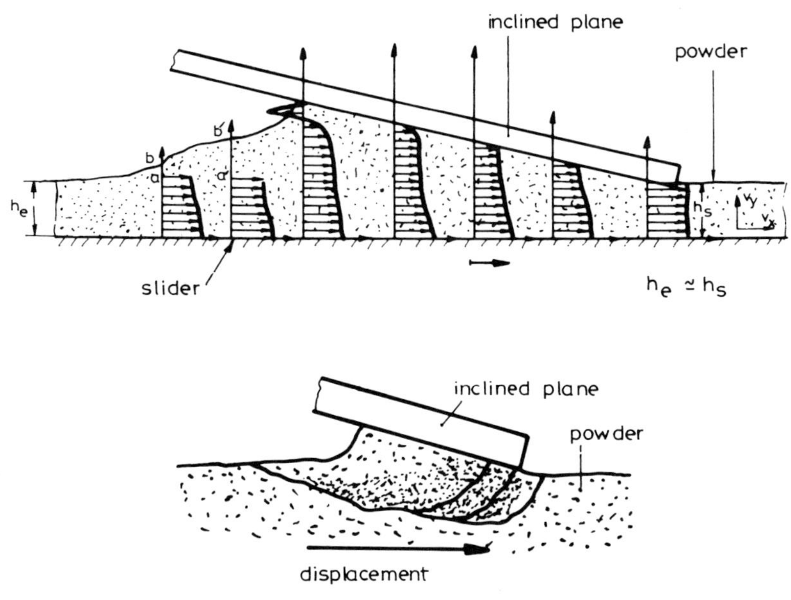

Fig. 9. Velocity Profiles and Rupture Fronts in Powder Beds (Reference 22).

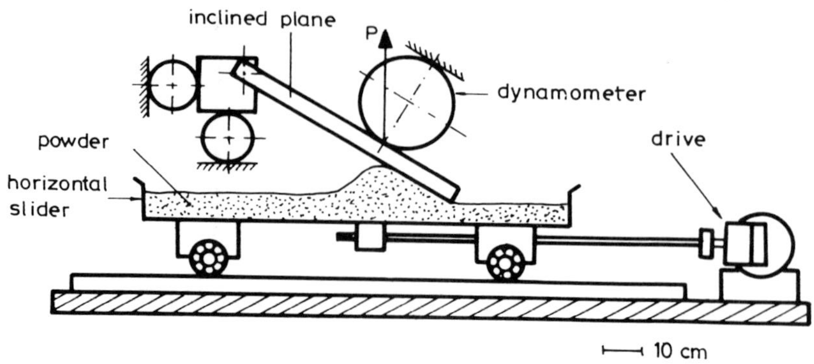

Fig. 10. Powder Load Carrying Test Device.

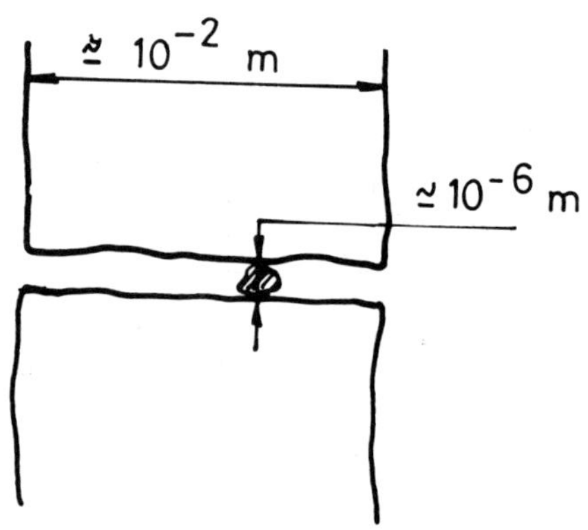

Fig. 11. Trapping.

TABLE I

Wear rates in cyclic tests

N.B.: Specific wear rates in M/MN are obtained by dividing u by the load N. Note that, instead of being constant as the theory suggests, they vary by a factor of 10 to 1 (ref. 13, p. 38)

Experimental results

N_0^* (N)	Max. ang. displacement (°)	$v_{av.}$ (m s^{-1})	f	W^{**}	θ (°C)	$\dot{u}^\dagger$
660	6	0.05	0.28	8.7	35	1.18
	12	0.09	0.32	19.9	43	0.74
	18	0.14	0.25	23.3	60	0.41
	24	0.19	0.19	23.5	54	0.665
			0.33	40.9	70	0.79
1000	6	0.05	0.21	9.9	37	1.9
	12	0.09	0.25	23.5	45	0.74
	18	0.14	0.21	29.6	60	0.4
	24	0.19	0.18	33.8	60	0.6
1330	6	0.05	0.185	11.6	35	3.55
	12	0.09	0.16	20	40	1.6
	18	0.14	0.13	24.4	50	0.4
	24	0.19	0.13	32.51	56	0.4
1750	12	0.09	0.11	18.1	50	0.6
	18	0.14	0.115	28.4	52	0.59
1930	24	0.19	0.1	36.3	55	0.82
			0.15	58.05	76	0.76

*$N = N_0(1 + \sin \omega t)$, cyclic load at 10 Hz.

**Heat input (W).

†Rate of wear ($\times 10^3$ μm m^{-1}).

The material used for the liner was Dacron weave cloth filled with Teflon.

TABLE II

Mechanical and materials' approach to three-body steady-state wear.

MECHANICS	MATERIALS
Boundary conditions { τ, μ	Surface physics { • surface energy • adhesion
Rheology { • pressure • temperature • shear-rate	Powders { • cohesion • specific surface • reactivity • adsorption

2. WORKSHOP SUMMARY

2. WORKSHOP SUMMARY

This summary represents the consolidated views of four panels during eight panel sessions, see **APPENDIX A** for the panels' membership and chairmenship for Sessions I and II. Session I of all panels addressed the question of modeling from the materials science vantage point. Session II of all panels addressed the question of modeling of friction and wear in general.

TWO LEVELS OF MODELS

Each level is envisaged to have a cascade of possible models. In other words, depending on the scope desired and within the realm of possibility based on current knowledge, there can be a series of models of varying complexity.

One of the two levels is the *Design-Oriented Level.* Referring to Fig. 1, for design purposes, the models should receive input on the one hand and produce output on the other hand. The dash and dot box is, for this level, viewed as a "black box." Often this view is a necessity. The black-box characteristic is experience-based. That is, it denotes a data bank for use by designers in predicting friction and wear in practical systems. More generally, associated with friction and wear, are such items as damage and transitions to be modeled. Data should be very well categorized, stating all parameters including shape, load, environment, materials, and treatments. The user would enter the data bank giving values of all parameters in the projected application. As such, at this level, the model is often in the form of a computer software. For complex processes modeling may be the construction of failure mechanism maps.

The second of the two levels is the *Fundamental Level.* Referring to Fig. 1, while the connection between output and input is again sought, ideally it is the quantification of the various relevant processes which must comprise the model or a cascade of models. As a step towards full quantification, models could be in the form of word pictures describing the mechanisms involved or the physical events which they entail. For complex processes, modeling may begin as empirical correlations.

DIMENSIONS OF TRIBOLOGICAL PROCESSES

It is important to recognize that there are several dimensions of tribological processes. It is useful to contemplate two such dimensions: (a) in the mechanism dimension, there are the likes of adhesion, elastic/plastic deformations, fracture, transfer, tribochemical interactions, and chemical reactions; (b) in the dimension of lubrication regimes, there are dry sliding, solid lubrication, boundary lubrication, mixed lubrication, fluid film lubrication, and more if warranted. This dimension is a systemic breakdown of a complex situation shown in Fig. 2. The global view of structural elements with tribological media interposed would pose a horrendous if not impossible task at modeling. Tribological media for the various lubrication subdivisions can be characterized as follows:

LUBRICATION REGIME	TRIBOLOGICAL MEDIA
Dry sliding	Mechanically-mixed third body made up of materials from structural elements (1) and (2) plus environment.
Solid lubrication	Solid lubricant and materials from structural elements (1) and (2) plus environment.
Boundary lubrication	Lubricant elements with materials from structural elements (1) and (2) plus environment.
Mixed lubrication	The above plus bulk lubricant.
Fluid film lubrication	Bulk lubricant.

It should be noted that, for completeness, fluid film lubrication regime was entered above. Fluid film lubrication, unlike the other regimes, has been most thoroughly modeled through the years. As such, within the context of the Workshop, modeling of fluid film regime is not being considered. Of course, wear due to transient conditions of start/stop is boundary lubrication and mixed lubrication which needs to be modeled. Also organo-metallics from oxidation and wear particles flow with the lubricant which cause accelerated wear; this class of phenomena needs to be modeled.

TWO TIERS OF MODELING AT THE FUNDAMENTAL LEVEL

The first tier of modeling desired is associated with the initial stage of friction and wear processes. That is the modeling of the formation of particulates from virgin engineering surfaces. Here important parameters would include surface topography and associated area of contact and surface tractions; surface chemistry and physics associated absorbed layers, cleanliness and adhesion; elastic and plastic deformation; void formation, crack nucleation and propagation; low cycle fatigue with associated deformation and fracture.

The second tier of modeling desired is associated with the behavior of debris once formed within the various tribological media. Debris should include chemical reaction products such as a single metallo-organic molecule. Here the important parameters would include adhesive properties of particulates and surfaces; cohesiveness of particulates; chemical composition of particulates; environmental factors; ambient and interface temperature; shape of particulates; type of geometry, loading, and motion. Statistical variations are to be noted. It should also be noted that in many cases the tribological media are protective and serve to limit wear of bodies ① and ②, while in other cases the media can, in fact, promote wear of the two surfaces. In short, morphology, rheology, and chemistry of the media must be characterized, and their relationship to the substrate and ambient liquid or gas must be established. While the above is general, it should be emphasized that for boundary lubrication and mixed lubrication the following are important: state of molecules in solution; adsorption of molecules on solid surfaces and particulate surfaces; reactions and film formation. Issues are complex non-equilibrium chemistry, dynamic processes, heterogeneous reactions, and reaction kinetics.

GENERAL COMMENTS

In light of the aforementioned, while most modeling, of necessity, will focus on pieces of the whole, an appreciation of the broad perspective must be kept in mind. Representation of models must be *communicative*. That is, the model gives the physical description. New methods such as computer graphics would help in illuminating the picture described. Modeling helps fundamental understanding, aids in new and needed measurement. In turn, the new understanding and new measurements made possible enhanced models.

The Workshop gave its strongest possible encouragement for modeling, be it at the *Fundamental Level* or at the *Design-Oriented Level*. Progress in the field of *Friction and Wear* depends on successes in enlightened modeling.

RECOMMENDATIONS

During the Workshop, the various panels considered what tribological processes should be modeled and recommendations were made. The selection was based on a variety of factors, the most important being need for a clearer understanding, sufficient data to evaluate the models and the lack of alternative methods to obtain the desired information. The following were suggested:

- Friction Model of the Boundary Lubrication

  Such a model would be important to energy conservation and to the proper selection of material/lubricant combinations. It would also give a better understanding of the separate roles of solid to solid contact and film formation. The model should include both effects as well as the control parameters that determine which effect controls frictional behavior under a given set of circumstances. Variables to be included are load, velocity, temperature, as well as material and lubricant parameters and conjunction geometry.

- Initial Formation of First Wear Fragment

  Wear processes are very complicated and often interact. On a fundamental level it is important to determine the independent, unique particle removal processes (i.e. plastic deformation, fatigue, cutting, fracture, transfer, cutting diffusion, melting, etc.) and to relate the size of the particle removed to the conjunction geometry and operating parameters of load, real contact area, temperature, and time and the material properties. Under effective lubrication conditions, most material is probably removed by chemical actions (i.e. adsorption of like molecules, reaction with surface, desorption of reaction product).

Friction and Wear of Well Characterized Surface Films

It is extremely difficult to characterize surface films because of the uncertainties concerning composition, structure, and properties under sliding conditions. A more useful approach is to define the tribological behavior which would be expected from films with well defined rheological properties. From these well defined properties, a battery of realistic rheological constitutive equations needs to be developed. Models should include both slip and wear submodels. Outputs should include wear rate as a function of film thickness and friction as a function of velocity and temperature.

Global and Local Stress Fields

Much work has been carried out in this area, but much more is needed, particularly under dynamic conditions which include the effects of temperature and wear. Also, large deformation and plastic deformation analyses await added attention.

Surface Temperature

Very little can be accomplished in modeling friction and wear processes until surface temperatures can be predicted. Models should include thermoelastic effects, film effects, and changing areas of contact and surface reactions. Much experimental data is needed as well as novel experimental methods.

Wear Behavior of Mechanically Mixed Surface Layers

A large amount of experimental work has been carried out to define the nature of the surface layers which result from unlubricated sliding. The next step is to define wear models which correlate with the observed wear behavior.

In concert with the modeling at the fundamental level, there is the need to model at the design-oriented levels and their interconnection.

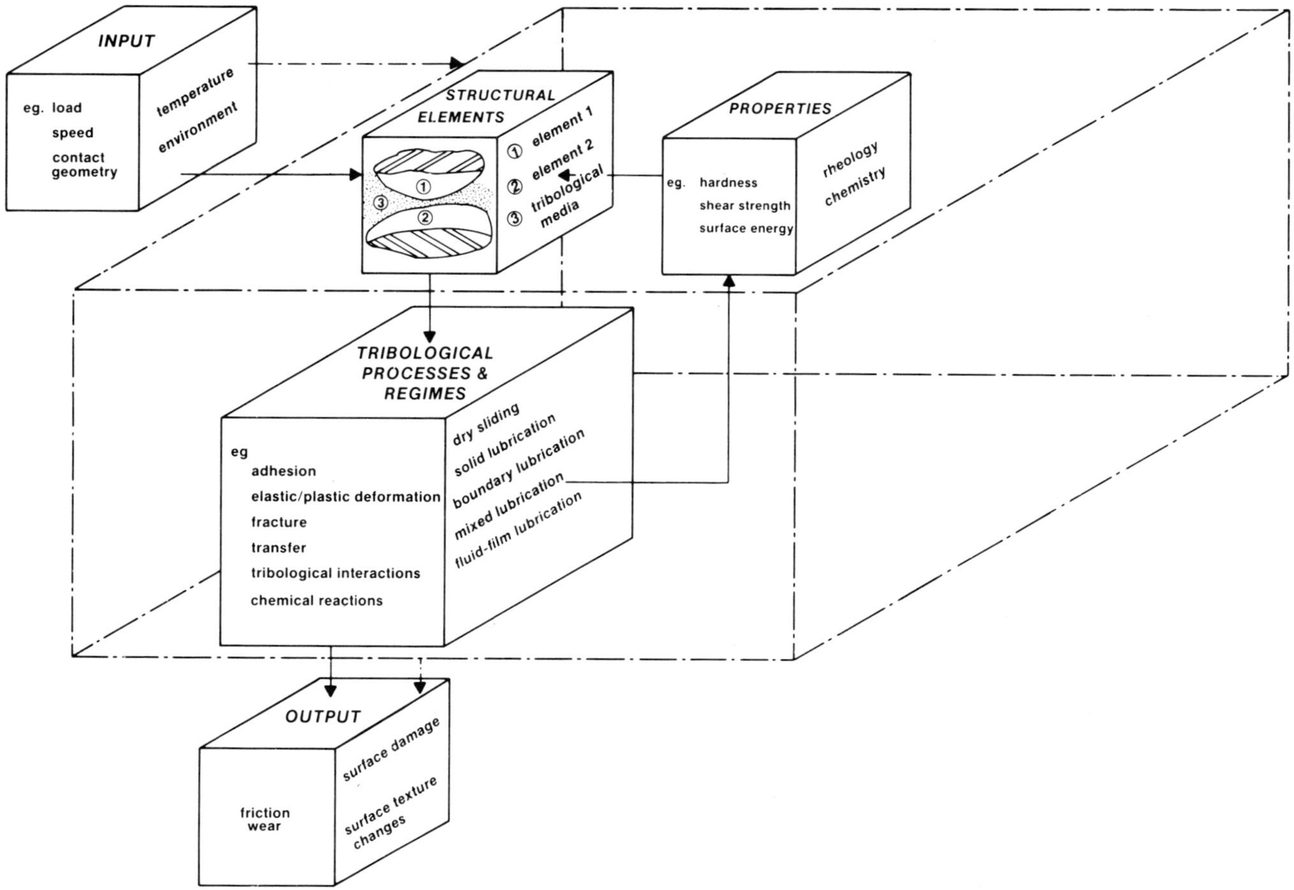

Fig. 1. Two paths from INPUT to OUTPUT: (a) INPUT to OUTPUT via dash and dot "black box;" (b) INPUT to OUTPUT via the solid-lined boxes.

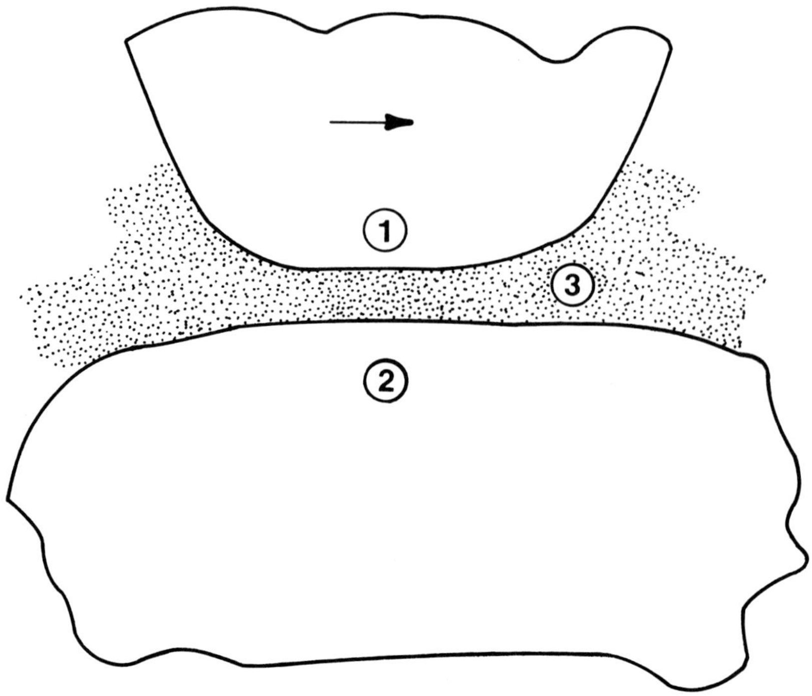

Fig. 2. Global view of structural elements in relative motion. (1) and (2) are the elements; (3) is the tribological media interposed between the elements.

APPENDIX A. MEMBERSHIP OF PANELS

There were four panels, each meeting in two sessions, to address the question of modeling from the materials science vantage point, and the question of modeling of friction and wear, respectively. The panels were:

Panel A. Gary L. Anderson, Army Research Office
Herbert S. Cheng (2), Northwestern University
Richard S. Fein (1), Richard S. Fein Associates and Columbia University
Morton B. Friedman, Columbia University
Jorn Larsen-Basse, Georgia Institute of Technology
Richard Nicholson, Rolls-Royce plc, United Kingdom
Glenn K. Rightmire, Columbia University

Panel B. David A. Dalman, The Dow Chemical Company
Said Jahanmir (2), National Science Foundation
Coda H. T. Pan, Columbia University
S. Y. Poon, Inter Technology Research Ltd., United Kingdom
Vishwanath Prasad, Columbia University
S. Ramalingam (1), University of Minnesota
John A. Tichy, Rensselaer Polytechnic Institute

Panel C. Norman S. Eiss, Jr., Virginia Polytechnic Institute and State University
Traugott E. Fischer, Stevens Institute of Technology
Donald G. Flom, General Electric Company
Maurice Godet, Institut National des Sciences Appliquees de Lyon, France
Sydney A. Karpe, David W. Taylor Naval Ship Research and Development Center
Francis E. Kennedy, Jr. (2), Dartmouth College
Richard C. Rosenberg, General Motors Corporation
Ponisseril Somasundaran, Columbia University
James J. Wert (1), Vanderbilt University
Julian J. Wu, U.S. Army Research, Development and Standardization Group, United Kingdom

Panel D. Frederick F. Ling, Rensselaer Polytechnic Institute and Columbia University
Fred Nichols, Argonne National Laboratory
Marshall B. Peterson, Wear Sciences Corporation
David A. Rigney, Ohio State University
Iradj G. Tadjbakhsh, Rensselaer Polytechnic Institute and U.S. Army Research Office
Lavern D. Wedeven (2), SKF Tribonetics
Ward O. Winer (1), Georgia Institute of Technology

(1) Chairman of first session; (2) Chairman of second session.

3. CONTRIBUTED BRIEF POSITION PAPERS

3.1 TRIBOLOGY MODELING

Donald G. Flom

The first step in approaching the analytical modeling of tribology requires that at least two questions be asked:

(1) What is the end objective of the modeling?

(2) Where are advances in tribology most needed?

My own answer to the first question is that a model is useless unless it is predictive. The offering of a plausible hypothesis to explain past experimental results or practical observations is not satisfactory unless it can point the way to new experiments or new approaches. It follows that modeling and the conduct of experiments must go hand in hand; they must feed on each other and thereby ensure that forward movement of the technology is nurtured. Modern instrumental and diagnostic techniques can help here because many events can now be monitored in real time rather than after the fact. For example, it is possible to videotape the rubbing and cutting of solids inside the scanning electron microscope, a situation vastly more revealing than looking at wear particles after they have been formed. Also, with sensitive vibration transducers one can listen to what is happening when solids are in contact and in relative motion. The interpretation of these signals may not always be easy but here again modern diagnostic techniques enable literally thousands of data points to be analyzed in a variety of different ways.

An answer to the second question, i.e. where are advances in tribology most needed, will depend on a person's parochial viewpoint and where his own interests lie. My own answer is that advances are needed at the extremes of environmental conditions. For example, we need increased tribological understanding in metal working, metal cutting, and associated processes. These processes usually entail high loads, high temperatures, high strains and high rates of strain. An example of a clue to the problem is that the coefficient of friction is very rarely measured directly in connection with any of these processes; it is inferred from other measurements. Also, satisfactory ways of simulating metal cutting and metal working processes are difficult to envision.

Another example of a need for tribological advances lies in the aerospace industry. The extreme environments of space make this quest obvious; unfortunately, the lack of commercial applications in space preclude this subject as a high priority item in industry. Closer to home, however, is the need for advances in tribology in connection with aircraft. Even slight increases in efficiency can be a tremendous boon to the makers of aircraft propulsion systems. The potential use of new materials such as ceramics, cermets, and new alloys makes this need especially urgent.

So where does this leave analytical modeling of tribology? The answer is that the modeling can serve a useful purpose if it forces the investigator to think carefully about his experimental results and observations, and particularly so if these do not agree with the model. By definition and also in practice, inventions stem more often from unanticipated results than they do from agreement with what was expected. When the prospects of coming up with new materials are dim, a new design may be called for. For example, if a wear-resistant material does not appear to exist for a given application, then a possible approach is to design a new system with increased tolerance to wear. This was in fact done in machining with the invention of the ledge cutting tool. There is no reason why similar considerations cannot be given to tribological systems.

3.2 MODELING OF DRY AND BOUNDARY LUBRICATED CONTACTS AS SEEN FROM THE HYDRODYNAMICIST'S PERSPECTIVE

John A. Tichy

A commonly held view among those reasonably familiar with tribology is that the modeling of a hydrodynamic thick-film contact, and even that of an elastohydrodynamic contact, is a "closed book"; whereas the modeling of a much thinner film, either boundary lubricated or nonlubricated, is effectively impossible.

We are fond of using the word "interdisciplinary" to describe our field, but too often we divide tribology neatly in two: hydrodynamics, and everything else (wear, boundary lubrication, etc.). The understanding of many problem areas which involve both features (e.g., piston ring lubrication, magnetic media lubrication, etc.) is probably proceeding more slowly due to our compartmentalized thinking. At tribology conferences there is relatively little discussion between the two factions and investigators in differing domains do not even know one another. I do not think this is a good state of affairs and it redounds to the detriment of our field.

The opinions in this brief position paper are restricted to the mechanical, as opposed to the materials science, point of view. The writer is well aware of the dangers of such a narrow outlook but is not sufficiently well-versed in materials science to comment intelligently. Investigators in the two camps (hydrodynamics vs. thin films) often do not think highly of one another. Many boundary lubrication experts feel that the century-old Reynolds equation is the be-all and end-all of hydrodynamic lubrication and is thoroughly understood, irrelevant, and overly analyzed. Many of my colleagues in hydrodynamics feel that the understanding of boundary lubrication is in such a primitive state that only simplistic, ad hoc, irrational models are available, and these only work over a very narrow range of highly idealized conditions.

The thesis of this brief paper is that the view of neither side is very accurate. At this conference, I have discovered that the situation in boundary and dry lubrication is not nearly so bleak as that described above. Many investigators have posed and solved boundary value problems using various plasticity models at a level of rigor which would please the most hard-nosed fluid mechanist, for example the work of Kennedy, Ohmae, Suh and others.

Perhaps I can enlighten the typical reader (presumably a thin-film expert) that hydrodynamic lubrication does not begin and end with the Reynolds equation. There are so many assumptions inherent in the Reynolds equation that it applies without modification only to a fairly narrow subset of hydrodynamics problems. In most applications, if thorough experimental verification were attempted I suspect that design predictions based on the Reynolds equation would be far afield. Fortunately, in most steady applications the design process is very forgiving and errors seem to fall on the conservative side.

Some "exceptions" to the Reynolds equation in hydrodynamics are now listed. Thermohydrodynamic behavior cannot, in principle or in practice, be adequately handled by the Reynolds equation. Real lubricants are strongly non-Newtonian, the behavior of which is not tractable by the conventional theory. In high-speed, lightweight machinery, such as that using composite materials, fluid inertia or turbulence will come into play. Other complications are due to roughness, non-continuum (low Knudson number) effects in gas bearings (e.g. computer heads), film rupture, particulate contamination, etc. There exists a vigorous research effort in hydrodynamics which is international in scope (note the Japanese contributions to hydrodynamics sessions at ASME-ASLE Tribology Conferences).

I have been pleasantly surprised and impressed by the work of Godet and his colleagues concerned with the "third body" models. The parallels to liquid lubrication are obvious and point to an approach which models the powdery third body through an appropriate constitutive equation (although not that of the Newtonian fluid). I have heard the criticism from Jahanmir that Godet's approach, and the approaches of many others who study dry lubrication, are not realistic in that their wear process is much more rapid and destructive than that in actual applications. This acceleration of wear is a device to facilitate research - real wear processes are far too slow to be efficiently studied. In most lubricant contacts, wear particles are created only rarely on an apparently random basis due to asperity interaction, fatigue, etc.

This brings me to a final comment. There is a relatively new branch of applied mathematics which is concerned with apparently random processes (such as wear events) that have deterministic underpinnings. This branch of mathematics generally carries the name of "chaos" but other key words are instability, phase plane, limit cycle, strange attractor and fractal. The basic idea is that a chain of events (such as the action of billiard balls on a table) which is entirely deterministic will become random (i.e. unpredictable) due to exponential growth in any minor variation at the outset (the initial "break" of tightly packed billiard balls). I suspect some ideas from this new field can be fruitfully applied to analysis of the wear process, much as it has been recently applied to the turbulent flow process.

3.3 ASSESSING RESEARCH NEEDS IN TRIBOLOGY

Julian J. Wu

The whole picture of lubrication technology looks somewhat like a jigsaw puzzle which has pieces tabulated in Table I. For a given machine we wish to know its expected service life for a given lubricant as a simple example. This picture can get very complicated; for economical and/or critical operations, many factors must be considered and understood. Some of them will be mentioned here.

A. Design Parameters:

- A-1. Material Systems: Machine (or biomedical) parts and lubricants
  Metals, alloys, composites ceramics; liquid and solid lubricants
  Chemistry: Material stability, degradation, effect of additives
- A-2. Surface Characteristics:
  Surface roughness
  Size and distribution of asperities (microstructure of surfaces)
  Fractal dimensions as means of surface characterizations
  Effect of scuffing
- A-3. Loading Characteristics:
  Sliding contact
  Rolling contact
  Combined rolling/sliding
  High load/high speed
  Fatigue
- A-4. Thermal Loads:
  Ambient
  Operating
  Thermal shock
- A-5. Geometric Characteristics:
  Journal/bearing
  Piston/cylinder
  Gear/gear
  Sliding planes

B. Correlation Techniques:

- B-1. Thermodynamics
- B-2. Constitutive Modelling
- B-3. Experimental Correlation:
  Various Test Procedures
- B-4. Dynamics Problems
  - Partial Differential Equations
- B-5. All Interrelations

C. Performance Parameters:

C-1. Service Life

C-2. High Efficiency (Low Friction)

C-3. Service Load/Service Temperature

C-4. Maintenance Interval

Thus, to solve a problem in lubrication is to complete such a jigsaw puzzle. Any missing links represent areas of research needs. For such a complicated system of correlations, one naturally thinks of building "expert systems" which store all of the available information (parameters, correlations, etc.), make necessary computations and provide optimal design solutions subject to given constraints, all with the speed of a computer.

Finally, such expert systems are not unique. Simple ones should be built at the beginning. They should also be dynamic in the sense that they can be updated easily with new research findings.

Table I. **SUNDRY FACTORS IN LUBRICATION TECHNOLOGY**

A.	Design Parameters	B.	Correlation Techniques (Parameters)	C.	Performance Parameters
A-1.	Material System Machine Parts/Lubricants	B-1.	Thermodynamics	C-1.	Service Life (Low Wear)
A-2.	Surface Characteristics	B-2.	Constitutive Modelling	C-2.	Low Friction (High Efficiency)
A-3.	Load/Speed	B-3.	Experimental Correlations	C-3.	Service Load/Temperature
A-4.	Thermal Load	B-4.	Dynamics Problems - Partial Differential Equations	C-4.	Maintenance Interval
A-5.	Geometric Characteristics	B-5.	All Interrelations		

3.4 AMONTONS AND COULOMB, FRICTION'S FOUNDING FATHERS*

Herbert Deresiewicz

As is now generally known, the idea that frictional force is proportional to normal force stems from a paper by Guillaume Amontons [1], published in the *Mémoires de Mathématique et de Physique* of the Académie Royale des Sciences in Paris for 1699, in the very first volume following renewal of the Academy's royal charter. Presented on December 19 of that year, it was entitled "On the resistance caused in machines, by the friction of their component parts as well as by stiffness of the cords used there, and the way to calculate the one and the other." I have inspected the paper in a second edition of the *Mémoires* published in Amsterdam in 1734 [2]. But rather than giving you my translation, I prefer to preserve some of the flavor of that time by quoting the translation in the abridgement of the *Mémoires* published in 1742 by two English academics [3].

At the very outset of his paper Amontons announces that "...we found

1. The resistance caused by friction increases and diminishes only in proportion to the greater or less pressures;

2. the resistance caused by the frictions is nearly the same in iron, copper, lead, and wood, let them be varied how you will, when these substances are anointed with old lard;

3. this resistance is nearly equal to one third of the pressure; to which must be added a fourth, the resistances are between themselves in a ratio compounded of the weights or pressures of the parts which rub, of the times, and of the velocities of their motion."

Indeed, six months earlier, on June 20, Amontons had read before the same Academy a paper entitled "A handy way of substituting the action of fire, instead of the force of men and horses, to move machines," in which he said, in passing, "... it is an error to think that the frictions in machines increase or diminish in proportion as the parts which rub are more or less extended." [4]

The general reaction was one of essential acceptance of the proportionality of the frictional to the normal force, mild skepticism concerning the almost universal value of 1/3 for the coefficient, and evident disbelief of the independence of the force of the area in contact.

The author of the *Histoire de l'Academie*, an epitome of and commentary on all that took place at the sessions, certainly pulls no punches. Of the assertion in the June paper that the force is independent of the contact area he says [5], "Cette nouveauté causa quelque étonnement a l'Académie." Thereupon Philippe de la Hire, now remembered primarily for his theory of arches and for his skillful use of

*In welcoming the Workshop attendees to Columbia, Professor Deresiewicz gave this historical account (with minor modifications).

graphical statics, rushed off to carry out his own experiment ["M. de la Hire consulta aussi-tôt l'expérience"] [5] with a view to shedding more light on the matter.

Amontons (whom his eulogy in the *Histoire* of 1705 [6] praises as having had "a singular gift for experimentation, fine and felicitous ideas, ..., great dexterity for execution, and one believed seeing revived in him Mr. Mariotte, so celebrated for the same talents") had attributed the presence of friction to surface roughness, the frictional resistance being, in the case of rigid surfaces, the force required to lift them over the asperities; or, if the surfaces are supposed capable of elasticity, the force needed to surmount the stiffness of the springs. But besides these possible mechanisms de la Hire suggested that, if the surface roughness were to wear away, the broken particles would contribute to the friction in some proportion to the area. Moreover, it was known that a pair of highly polished surfaces, once pressed together, were difficult to separate. For this phenomenon de la Hire gave what would now be considered a rather fanciful explanation, but space limitations preclude me from going into that.

The unnamed author of the *Histoire* (identified by the authors of the *Philosophical History* as "the celebrated M. Fontenelle," the Academy's perpetual secretary) adds that when "the pressure of contact is very slight and the number of parts to be rubbed very great, ... it must then happen that the friction sensibly follows the proportion of the surfaces." He concludes, "This sort of metaphysics ...may serve to give a more complete and perfect idea of the frictions than that which might be drawn solely from experiments."

Amontons' proportional law was widely accepted in papers of the 18th century, among them those by the talented and critical Antoine Parent and by that most rational of rational mechanicians, the great Leonhard Euler, although suggestions were voiced, however vaguely, that the *cause* of friction be sought in cohesion of the surfaces.

Charles Augustin Coulomb, possibly the first to apply the calculus to engineering problems, came to our subject in 1779, when the Academy proposed for the second time as the theme of its now doubled prize for 1781 (the prize in the 1779 competition on the same topic having been withheld) the solution of problems of friction of sliding and rolling surfaces, the resistance to bending in cords, and the application to simple machines. A military engineer at the time, Coulomb was able to use the facilities of a naval shipyard for his experiments. From these experiments, and those of his predecessors, he developed a general theory of friction and a series of empirical formulas. His lengthy memoir, "Théorie des machines simples," won him the first prize, the second one of his career, and election to the Academy shortly thereafter. The work was printed in the *Mémoires...par divers savans*, 1785, in which it occupies fully 172 pages, the first 99 devoted to sliding friction [7].

Coulomb acknowledges Amontons as "the first author who tried to evaluate friction and stiffness of cords in the calculation of machines," but notes that the highly regarded Dutch physicist, Pieter van Musschenbroek, and others found that "the extent of the surfaces had influence," and that "neither one nor the other searched to determine the relation that could exist between" "the friction of a moving body and that of a body one wished to start [moving] from a state of rest."

In short, what Coulomb undertook and carried out was a comprehensive program of wide-ranging experimentation in sliding and rolling friction that entailed repeating many tests and investigating numerous parameters. He found the most important parameters to be: the nature of materials in contact and their coating; the surface area; the normal force; the time of contact before motion begins (time of repose); and the relative velocity. In his analytical formulation he used an additive constant term to account for a cohesion or surface film effect, although he found this effect, in general, to be less than 5% of the total.

Finally, he gave an explanation of why, although the coefficient of friction was pretty much the same for static and dynamic conditions in metals, it varied with time of repose and the velocity of motion for fibrous materials, this explanation based on time-dependent deformation of the fibers. In contrast with Amontons' work, whose rules could not be borne out by his experimental data because the latter contained vast ranges of extrapolation, Coulomb had a firm experimental basis for his conclusions.

A capsule summary of our two protagonists can be gleaned from the words of contemporaries, to be found in their respective memorials:

"[Amontons] gave in our meetings his theory of frictions, which clarified so important a matter in mechanics, and until then so obscure" [6].

"...one sees [Coulomb] at first attentively questioning nature; seizing and ascertaining important facts; then searching in rational mechanics for the most appropriate formulas to tie together the isolated facts; carrying out new experiments and varying them in all appropriate ways to obtain and determine the constants in his formulas and the quantities that could vary in accordance with the nature of the substances [in the] experiments" [8].

One cannot leave the history of friction studies without at least touching upon the oft encountered assertion crediting early thinking on the subject to Leonardo da Vinci. Dowson [9] had devoted considerable space to the tribological writings of this individual (whom a noted historian of science characterized as "endowed with insatiable curiosity" [10], and who perhaps more than anyone in our civilization embodied the virtues implied by the epithet "Renaissance man"), adducing firm evidence for the contention that Leonardo had anticipated Amontons in stating that friction is proportional to the normal load and independent of the contact area. To be sure, the nagging question remains about how much, if indeed any, influence Leonardo's scientific thought could have exerted on his contemporaries and on succeeding generations if his works, consigned to pretty much inaccessible manuscript notebooks, remained unpublished for some 400 years.

Here we confine ourselves to an entry in a notebook begun in September of 1508. Discussing various aspects of the Sun, Leonardo considers "Whether the friction of the heavens makes a sound or no," and in short order argues himself into an interesting quandary:

"...in the case of smooth bodies the friction does not create sound, and it would happen in a similar manner that there would be no sound in the contact or friction of the heavens. And if these heavens are not smooth at the contact of their friction it follows that they are full of lumps and rough, and therefore their contact is not continuous, and if this is the case the vacuum is produced, which it has been concluded does not exist in nature. We arrive therefore at the conclusion that the friction would have rubbed away the boundaries of each heaven,...and then there would not be friction any more, and the sound would cease, and the dancers would stop..." [11].

Thus, we have learned that friction has its celestial as well as its terrestrial role, a fact that should serve as yet another incentive for continued vigorous pursuit of tribological researches.

NOTES

1. This has not always been accepted or even acknowledged in our century. For example, a standard treatise, the *Cours de Mécanique* by Leon Lecornu (Paris, 1914, vol. 1, p.473), states, "Les recherches expérimentales de Coulomb, puis du général Morin, ont établi que le coefficient f [the coefficient of friction] est indépendant de la force de pression...." Nor is that a unique case; Paul Appell, in his justly renowned *Traité de Mécanique Rationelle* (6th ed., Paris, 1941, vol. 1, p.315), expressed it in much the same way: "Les premières expériences sur le frottement de glissement, faites par Coulomb...."

2. *Histoire de l'Academie Royale des Sciences Année M.DC.XCIX*. Avec les *Mémoires de Mathématique et de Physique*, pour la même Année. Tirés des Registres de cette Académie. Seconde Edition, revue, corrigée & augmentée. A Amsterdam. Chez Pierre Mortier M.DCCXXXIV. *Mémoires*, pp. 259-286 + 5 tables.

3. *The Philosophical History and Memoirs of the Royal Academy of Sciences at Paris*: or, An Abridgement of all the Papers relating to Natural Philosophy, which have been published by the Members of that Illustrious Society, from the Year 1699 to 1720. With many Curious Observations relating to the Natural History and Anatomy of Animals, &c. Illustrated with Copper-Plates. The Whole Translated and Abridged. By John Martyn and Ephraim Chambers. In Five Volumes. London: Printed for John and Paul Knapton and John Nourse M.DCC.XLII, vol. 1, pp.112-131.

4. Ibid., p. 77. Also, note [2], *Mémoires*, p.172.

5. Note [2], *Histoire*, p.138.

6. "Eloge de M. Amontons," *Histoire de l'Academie Royale des Sciences Année MDCCV*. Avec les *Memoires de Mathematique et de Physique*, pour la même Année, Tirez des Registres de cette Academie. A Amsterdam. Chez Gerard Kuyper MDCCVII. *Histoire*, pp.189-194.

7. "Théorie des Machines Simple. En ayant égard au frottement de leurs parties, et a la roideur des Cordages," C. Coulomb, *Mémoires de Mathématique et de Physique,* Présentés a l'Académie Royale des Sciences, par divers Savans, et lus dans ses Assemblées. Paris, vol. X, 1785, pp.161-332 + 5 plates.

8. "Éloge historique de M. Coulomb," Delambre, *Mémoires de la classe des sciences mathématiques et physiques de l'Institut National de France,* deuxième semestre de 1806, vol. 7, 2nd part, Paris, 1807. *Histoire*, pp.206-223.

9. *History of Tribology*, D. Dowson, Longman, London and New York (1979).

10. "Renaissance Physics," Pierre Duhem, in *Toward Modern Science, II,* Robert M. Palter, editor, The Noonday Press, New York, 117 (1961).

11. *The Notebooks of Leonardo da Vinci*, E. MacCurdy, *I*, George Braziller, New York, 282 (1958).

4. CONTRIBUTED BRIEF SCIENTIFIC PAPERS

4.1 ON USE OF SURFACE DEFORMATION MODELS TO PREDICT TRIBOLOGICAL BEHAVIOR

Richard S. Fein

Friction and, to a greater extent, wear and surface damage from strong inter-surface adhesions or plastic flow depend critically on deformation (of bearing materials). These manifestations of tribological behavior also depend on the other physical processes and on chemical processes and how these interact with system variables (including operating variables) and among themselves. Thus, deformation models can be useful for explaining, predicting or correlating tribological behavior only as a part of a model including the other significant processes for a particular wear regime(s).

Table I lists the system variables and physical and chemical processes that appear to be important to tribological behavior under boundary lubrication conditions [1]. Boundary lubrication (BL) occurs when the mean surface separation is smaller than the composite highest interacting asperity height or largest particles between the surfaces in the bearing load carrying zone (LCZ). All the processes, except for those dependent on the viscous liquid-like properties of the film separating the LCZ and asperity interaction (AI) zones, appear pertinent to solid-film lubricated and unlubricated sliding.

Friction, wear and surface damage often are extremely sensitive to operating conditions and bearing, lubricant and atmosphere composition. The sensitivity is demonstrated most forcefully by the effect of composition on the operational severity (e.g. combination of load and speed along with the other system variables) at which abrupt changes in the level of friction, wear and surface damage occur. Fig. 1 illustrates the regimes of BL commonly observed for ferrous surfaces lubricated with hydrocarbon-base oils [2,3]. Usually at least one each of mild wear (MW), damage-controlled wear (DCW) and surface-damage wear (SDW) regimes is observed.

The transitions between regimes, the levels of friction and wear coefficients within the regimes, and the nature of the surface damage in the SDW regime commonly depend on the entire set of system variables (including run-in represented by time history). All bearings, regardless of the mode of lubrication, seem to similarly exhibit different regimes with transitions between them.

Mechanisms of lubrication, friction and wear appear to be different in each of these regimes. Presumably the different mechanisms result from dominance of a different set of series and parallel pathways among the physical and chemical processes of Table I. Process dominance can change suddenly as exemplified by collapse of the entraining of high-modulus EHD films when too much load is supported by asperity interaction [4].

RECOMMENDATIONS

Modelers and experimentalists should locate the subject of their work in the vast area of bearing system variables and the regimes of tribological phenomena. Regime identification is an acute need which is not satisfied in much published work.

Modelers of deformation should include the other physical and chemical processes pertinent to the causes of deformation and the fate of deformed material if tribological phenomena are to be modeled (tribochemistry, abrasion, etc. should be substituted for deformation by those active in those fields).

Experimentalists should make their results useful for verifying models or as data sources for friction and wear coefficients by providing information on (1) the extent to which friction and wear coefficients derived from the data represent steady-state conditions (i.e. run-in has been completed) and (2) the existence and nature of any surface damage.

Experimentalists and modelers always should be aware of the facts that (1) compositional analyses of bearing surfaces after sliding never completely determine all chemical elements and their chemical states and (2) physical properties of bearing surface films after sliding can be substantially different from bulk properties of similar material.

REFERENCES

1. "A Perspective on Boundary Lubrication," R. S. Fein, *Industrial and Engineering Chemistry Fundamentals*, *25*, 518-524 (1986).
2. "Failure Criteria in Thin Film Lubrication: Investigation of the Different Stages of Film Failure," H. Czichos, *WEAR*, *36*, 13-17 (1976).
3. "Failure Mechanisms in Sliding Lubricated Concentrated Contacts," A. W. J. deGee, A. Begelinger and A. Salamon, *11th Leeds-Lyon Symposium*, IPC Science and Technology Press, Guildford, UK (1984).
4. "The Failure of Elastohydrodynamic Lubrication of Circumferentially Ground Discs," A. Dyson, *Proceedings Institution of Mechanical Engineers (London)*, *190*, 52/76, 699-711 (1976).

Table I. SYSTEM VARIABLES AND PHYSICAL AND CHEMICAL PROCESSES

System Variable

- Bearing
- Geometry
    - Convergence
    - LCZ width
    - LCZ length
    - Surface motions
        - Slide/sweep
    - Out-of-LCZ/LCZ
    - Heat paths
- Roughness
    - Height
        - Tip radius
        - Valley volume
    - Directionality
- Material
    - Bearing
    - Lubricant
- Atmosphere
- Contaminant
- Lubricant supply
    - Volume
    - Application
    - Filtration
    - Temperature
    - Oxygen, water
- Load
    - Pressure
- Speed
    - Mean velocity
    - Sliding velocity
    - Sweep velocities
        - LCZ time/cycle
        - Non-LCZ time/cycle
- Temperature
    - LCZ film
    - LCZ substrates
    - LCZ entrance
    - Non-LCZ substrate
- Time
    - Sliding distance
    - Non-LCZ reaction

Physical Process

- Displacement\
    - Bulk fluid
    - Liquid-like film*
    - Particles*
- Entrainment\
    - Bulk fluid
    - Liquid BL film
    - Particles
    - Oxygen, water
- Shear entrained film\
    - Strain
    - Stress
    - Move particles
    - Deform particles
    - Heat
- Plowing
    - Large part./solid substrate deformation
    - Incr. roughness+
    - Cut□**
    - Extend BL film
- Deformation
    - Damage accumulation
    - Roughness+
- Fracture
    - Particle□
    - Nascent surface
    - Roughness+
- Adhesion
    - Transfer□*
    - Roughness+

Chemical Process

- LCZ reactants
    - Bulk fluid
    - Liquid-like film
    - Solid-like film
    - Oxygen, water
    - Substrates
    - Particles
- LCZ integral reactor
    - AI sub-reactors
- Reaction mechanisms
    - Radical generation
        - Thermal
        - Shear
    - Catalysis
        - Transition metal ion
        - AW-EP additives
        - Particle surface
        - Nascent surface
    - Oxidation
        - Peroxide generation
        - Substrate oxidation
        - Decomposition
        - Polymerization
    - Reduction
        - Decomposition
        - Hydrogen generation
    - Hydrolysis
    - Naturalization
- Diffusion
- Adsorption
    - Feed reactants
    - Reaction products
- Corrosion
    - Oxidation
    - Neutralization
    - Electrochemical
- Solubilization
    - Dispersion□*
    - Desorption□*
    - Adhesion
- Melting

\ Hydrodynamic Process
□ Substrate Detachment Process
* Wear Process (Removal of Detached Substrate)
+ Surface Damage Process

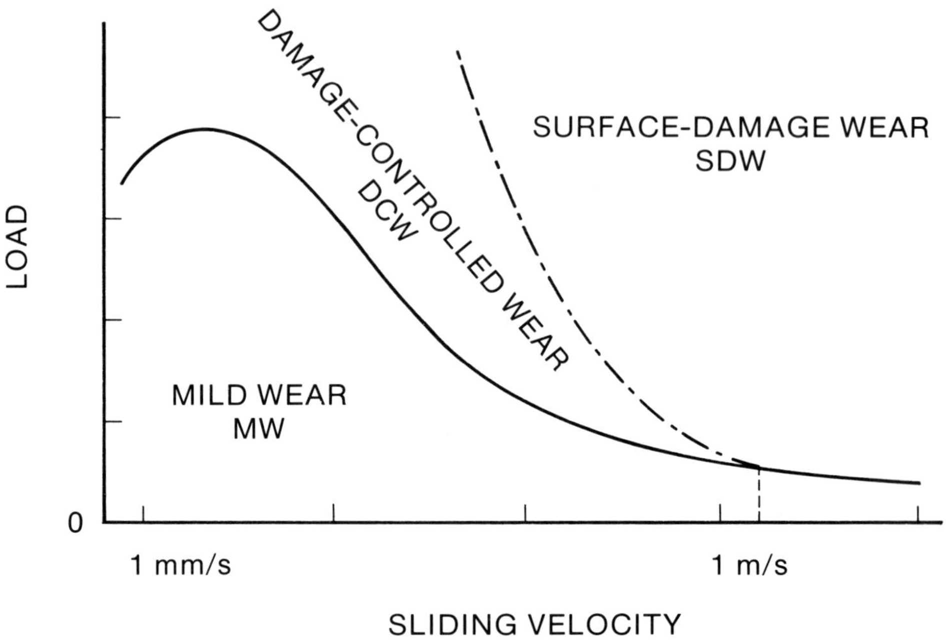

Fig. 1. Boundary Lubrication Regimes.

4.2 MODELLING TRIBOCHEMISTRY

Traugott E. Fischer

INTRODUCTION

The influence of chemistry on the sliding and rolling wear of materials has been known for a long time [1]. This influence is exploited in the formulation of lubricants where anti-wear additives react to deposit protective coatings on the rubbing surfaces and extreme pressure (E.P.) additives prevent the seizing of gears by controlled corrosion that prevents the welding of metallic contacts. Chemistry also manifests itself in fretting wear of bolted junctions where low amplitude rubbing causes accelerated oxidation, in the strong influence of environment on the wear of ceramics and in oxidative wear. A common feature of all these phenomena is the much higher rate of chemical reaction on the rubbing surfaces than on the adjacent areas of the same pieces of material. This interaction of friction with chemical reaction is called tribochemistry and is the subject of this analysis.

It is generally considered that the high temperatures generated by friction are responsible for the local enhancement of reaction rate in the action of anti-wear and extreme pressure additives, but large enhancements in reactivity occurring in fretting wear and in laboratory experiments where low sliding velocities and loads minimize frictional heating indicate that other mechanisms can accelerate chemical reactions as well.

Tribochemistry participates in a range of phenomena; it is responsible for changes in lubricant chemistry, it is prominent in ball milling and it modifies the wear of lubricated and unlubricated sliding bodies. Since we are concerned here with an approach to modelling rather than a survey, we shall restrict ourselves to oxidative wear of metals and to the chemical aspects of ceramic wear for no better reason than the author's personal experience. We shall consider first the acceleration of chemical reactions by friction and then the modification of wear rates and wear mechanisms by chemical reactions or their products.

TRIBOCHEMICAL REACTIONS

Before we can describe how friction accelerates chemical reactions of a solid with its environment, we must consider the reactions the solids undergo in the absence of sliding. Extensive research in surface science has now well established that oxygen and most other molecules adsorb readily on clean metal and ceramic surfaces and form strong chemical bonds with them. The slow step in a continued reaction is the diffusion of the reacting atoms or ions through the layer of reaction product. One knows, for instance, that the oxidation of iron will follow a parabolic law, in which the oxide layer thickness ξ increases as the square root of time τ:

$$\xi = c \sqrt{\tau} \tag{1}$$

at elevated temperature [2] and a logarithmic law

$$\xi = a \ln(t/t_0+1) \tag{2}$$

at room temperature [3]. The parabolic law corresponds to thermally activated diffusion of the oxygen anions and metallic cations through the oxide layer and the logarithmic law is the consequence of electromigration of the ions in the large electric field set up by the oxygen anions adsorbed on the oxide surface. These anions were formed by quantum-mechanical tunneling of electrons through the oxide. Nonoxide ceramics, such as silicon nitride (Si_3N_4) and silicon carbide (SiC) oxidize at high temperatures with parabolic kinetics. The oxidation of these materials is accelerated about one thousandfold by the presence of water vapor [4]. No measureable oxidation of these ceramics has been observed near room temperature.

When subjected to friction, steel oxidizes much more rapidly and the amount of material oxidized increases linearly with time at a rate that depends on sliding velocity [5]. With silicon nitride, friction leads to the formation of measurable quantities of silicon oxide and to relatively rapid dissolution of the material in liquid water even at room temperature [6].

Our task is to model the mechanisms by which the kinetics of the chemical reactions are modified by friction. We shall consider four such mechanisms, namely, frictional heating, frictional removal of oxide scale, friction-induced modifications in the structure of scale and substrate, and direct dynamic activation. We shall discuss frictional heating last because the existence of the other mechanisms will show that consideration of temperature alone leads to wrong conclusions.

The quasi-periodic removal of oxide scale by friction is a fundamental mechanism of oxidative wear. Not only does it increase the net oxidation rate, but it is responsible for the material removal that constitutes wear. It is convenient to approximate this phenomenon by taking an average time τ between successive scale removals and to assume that the oxide grows from thickness zero during that time according to some kinetic law. This mechanism leads to an oxide growth rate that is linear in time:

$$X(t) = \xi(\tau)\frac{t}{\tau} \tag{3}$$

where t is the total sliding time. If the oxidizing area is A = L/H, where L is the load and H the hardness of the steel, the volume of oxide grown and removed at each step is

$$\Delta V = \frac{L}{H}\,\xi(\tau) \tag{4}$$

and the total amount of material removed by oxidative wear is:

$$V(t) = \frac{L}{H}\,\xi(\tau)\,\frac{t}{\tau} \tag{5}$$

The average growing time τ depends on the kinetics of scale removal. Quinn [7] proposes that the oxide is removed every time it reaches a critical thickness ξ_c after having grown with the parabolic law. In this case, $\xi(\tau) = \xi_c$ and (1) gives $\tau = \xi_c^2/C^2$ so that (5) becomes

$$V_{Quin}(\tau) = \frac{LC^2}{H\xi_c} \times t \qquad (6)$$

Quinn's theory was derived for relatively high sliding velocities that cause an appreciable increase in temperature because of frictional heat. Since diffusion is thermally activated, the parabolic rate constant C takes an Arrhenius form:

$$\Delta V = \frac{LC_s^2 \exp(-2Q/kT)}{H\xi_c v} \times S \qquad (7)$$

where S is the sliding distance, v is the velocity and T is the surface temperature of the sliding bodies. Equation (7) is a simplified version of Quinn's formula which contains terms allowing for a real contact area divided into a number of spots. Much of the remaining modelling in Quinn's theory relates the surface temperature T with the load, the sliding velocities and details of the contact geometry [8]. Notice that in Quinn's theory, the wear rate depends on the velocity through the temperature that is generated by friction.

Sexton and Fischer [9] have designed their experiments to keep frictional temperature increases at an insignificant $\Delta T < 2K$ by using small loads and velocities. They observe that oxidative wear is linear in sliding distance and varies with velocity as:

$$\Delta V = \frac{LC}{H\sqrt{\lambda}} \times \frac{S}{\sqrt{v}} = \frac{LC}{H\sqrt{\lambda}} \times \sqrt{vt} = \frac{LC}{H\sqrt{\lambda}} \times \sqrt{St} \qquad (8)$$

where λ is an average sliding distance defined below and all other terms are as above. Equation (8) is obtained directly from (5) if one assumes that the oxide film is removed, whatever its thickness, every time the surfaces have rubbed over the average distance λ. The weakness of Equations (7) and (8) is that they contain an unknown parameter (ξ_c or λ). By prescribing a physically meaningful thickness ($\xi(\tau) > 1$ nm) for the film at removal, one can set a lower limit to λ(> 10 nm) and to the oxidation rate constant C in Equation (8). Comparison with experimental values of ΔV then leads to the conclusion that friction changes the growth law of the oxide film at room temperature from logarithmic to parabolic and causes the oxidation rate to be as large as it would be at 900K in the absence of friction. Thus, at least one further tribochemical mechanism is at work and increases the diffusion rate of reagents.

The previous finding signifies that during rubbing of the surfaces, diffusion becomes more rapid than electromigration. Diffusion is greatly facilitated by structural defects such as grain boundaries, dislocations, vacancies; it is well known also that large numbers of such defects are introduced by deformation in the surface of materials. We conclude that structural modification of the material, especially the creation of defects, constitutes another mechanism by which friction increases chemical reactions. The existence of a layer of material with highly excited defects is further demonstrated by such phenomena as triboluminescence and exoelectron emission.

Although we have not obtained direct evidence for its existence to date, we must keep in mind the possibility of a further mechanism of tribochemistry, namely the direct transfer of energy from mechanical friction to chemical reaction. Such a transfer is possibly involved in stress corrosion cracking; the very large stresses at the crack tip increase the interatomic distance, destabilize the bonding orbitals and stabilize the antibonding ones. This reduces the activation energy for a chemical reaction. This phenomenon could well be responsible for the increase in the wear of zirconia in humidity by stress corrosion cracking [10].

The temperature dependence of tribochemical wear presents an interesting and relatively complex problem. On the one hand, high temperatures activate the reaction, usually by activating the diffusion of reagents. On the other hand, they cause the annealing of the defects generated by friction and diminish the mechanically induced acceleration of diffusion, bringing the reaction rate closer to that which is obtained in the absence of friction. Thus an increase in temperature, either externally produced or generated by friction, may produce an increase or a decrease in tribochemical reaction rate. Heinicke and coworkers [11], for example, found the tribochemical reaction rates in their milling experiments to be independent of temperature.

MODIFICATION OF THE WEAR PROCESSES

Tribochemical reactions often cause large changes in the wear rates, increasing or decreasing them according to the sliding material and the environment. Changes in the friction coefficient can also occur, but there is no simple correlation between friction coefficient and wear rate; in zirconium oxide, for instance, stearic acid decreases the friction coefficient from $f = 0.4$ to $f = 0.12$ but produces a wear rate more than ten times larger than dry nitrogen does.

Tribochemical reactions change the wear of the original material by modifying the stress distribution, by influencing the material's response to the stress, and by removing reaction product. Purely mechanical wear usually produces rough contact surfaces leading to stress concentrations responsible for plastic deformation and fracture. Tribochemical reactions, which remove contacting asperities, produce smooth surfaces which allow lower maximum stresses at constant force.

Stresses are also distributed by surface layers of reaction products that are generally softer than the substrate. This decrease in the maximum value of the stresses imposed by an unchanging load suppresses the purely mechanical forms of wear that occur in inert environment and allow a milder form of wear to operate. The increase in actual contact area is responsible for the fact that the friction force remains often unchanged. Silicon nitride sliding in water, for instance, acquires a highly polished surface and a wear rate 100 times lower than when it slides in dry gases.

The modification of the material's response to stresses takes several forms. In the oxidative wear of steel, for instance, large incompatibility stresses weaken the steel-oxide interface and reduce wear by detachment of the thin oxide instead of deeper crack propagation. In oxide ceramics, notably zirconium oxide, water molecules and probably polar hydrocarbons weaken the grain boundaries and accelerate mechanical wear by intergranular fracture [10].

Modelling the friction coefficients is more difficult than modelling wear. We can invoke decreased adhesion, the corrosion of junctions that are forming and the formation of films with low shear modulus that are thin enough not to allow an increase in real contact area. The latter is perhaps the most amenable to quantitative description. Even if we do not succeed in establishing a model of the friction forces, their modification by tribochemistry must serve as an input into wear models since wear is normally a rapid function of the friction coefficient.

OVERALL MODEL OF TRIBOCHEMICAL WEAR

A description of tribochemical wear must incorporate the combined action of tribochemical reaction and wear mode. The quasi periodic removal of the scale transforms a given non-linear growth rate of the reaction into a linear wear rate that depends on the sliding velocity. The velocity dependence of the wear rate thus turns out to be the chief method to determine the reaction kinetics and, by extension, the chemical reaction path. When the velocity is high enough, frictional heat will generate high temperatures that will activate the chemical reaction.

In the case of ceramics, we have been able to show the influence of tribochemical reactions on the wear process and wear rates, but the experimental results were too complex to permit the elaboration of a simple model. Simple experiments that eliminate as many variables as possible will be necessary to identify the mechanisms at work, probably by measurement of simple dependence on sliding velocity or on temperature.

REFERENCES

1. "The Wear of Metals Under Unlubricated Conditions," J. F. Archard and W. Hirst, *Proceedings of the Royal Society (London), A236*, 397-410 (1956).
2. *Corrosion and Corrosion Control, An Introduction to Corrosion Science and Engineering*, 3rd edition, H. H. Uhlig and R. W. Revie, John Wiley and Sons, New York, 190-193 (1985).
3. "Die Geschwindigkeit der Einwirkung von Sauerstoff, Schwefelwasserstoff und Halogenen auf Metalle," G. Tammann and W. Koester, *Zeitschrift fuer Anorganische und Allgemeine Chemie, 123*, 196-224 (1925).

4. "Effect of Water Vapor on the Oxidation of Hot-Pressed Silicon Nitride and Silicon Carbide," S. C. Singhal, *Journal of the American Ceramics Society, 59*, 17 (1975).
5. "Surface Film Formation and Metallic Wear," W. Hirst and J. K. Lancaster, *Journal of Applied Physics, 27*, 1057-1065 (1956).
6. "Interaction of Tribochemistry and Microfracture in the Friction and Wear of Silicon Nitride," T. E. Fischer and H. Tomizawa, *WEAR, 96*, 17-30 (1984).
7. "The Effect of 'Hot Spot' Temperatures on the Unlubricated Wear of Steel," T. F. J. Quinn, *American Society of Lubrication Engineers Transactions, 10*, 158-160 (1967).
8. "Review of Oxidational Wear, Part II: Recent Developments and Future Trends in Oxidational Wear Research," T. F. J. Quinn, *Tribology International, 16*, 305-315 (1983).
9. "The Mild Wear of 52100 Steel," M. D. Sexton and T. E. Fischer, *WEAR, 96*, 17-30 (1984); "The Tribochemistry of Oxidative Wear," T. E. Fischer and M. D. Sexton, *Physical Chemistry of the Solid State: Applications to Metals and Their Compounds*, P. Lacombe, editor, Elsevier Science Publishers, Amsterdam, 97-108 (1984).
10. "Friction and Wear of Tough and Brittle Zirconia in Nitrogen, Air, Water, Hexadecane and Stearic Acid," T. E. Fischer, M. P. Anderson, S. Jahanmir and R. Salher, *Wear of Materials, Proceedings of the International Conference on Wear of Materials*, Houston, TX (1987), in press.
11. *Tribochemistry*, G. Heinicke, Carl Hanser Verlag, Munich (1984).

4.3 ROLE OF NANOSTRUCTURE OF ADSORBED LAYERS IN LUBRICATION

Ponisseril Somasundaran

Surfactants and polymers added to boundary lubrication systems adsorb on solid surfaces and form a protective film. Effectiveness of boundary lubricants has often been attributed to the adsorption affinity and the integrity of the adsorbed film. Such adsorption is influenced by additives introduced into the system to reduce thermal degradation, corrosion, sludge formation, foaming etc. There are many interactions that can take place between the additives and the surfactants and the base oil leading in addition to adsorption effects, a number of interfacial and colloidal phenomena such as micellization, precipitation and solubilization as well as flocculation of particulate matter in the fluid [1]. Clearly the adsorption and integrity of the adsorbed film will depend on the molecular geometry of the interacting species. For example, among various saturated mono-, bi-, and tricyclic aromatics, alkanes, and polyethers tested as lubricating fluids by Hentschel [2], the "irregularly shaped molecules" yielded high traction coefficients whereas molecules with a "thread-like shape and minimum structural subunits or functional groups" yielded low coefficients. Evidently ability of molecules to undergo intramolecular as well as intermolecular motions is important in maintaining the integrity of the lubricating films.

Geometry of the molecules in the adsorbed layer and the resultant structure of the adsorbed layer on a molecular scale is fundamental to the lubrication phenomena and a knowledge of the molecular mechanism on that scale is essential for any major technological leaps in this area. However, while there has been a significant amount of work on a *macroscopic* scale to study the thickness of the film, its rheology, thermal conductivity, oxidation [3] ability and corrosion there has been only a very limited effort to determine adsorption properties [4] and no studies, to the author's knowledge, to understand the molecular structure of the adsorbed film.

Conventionally adsorption in investigated by macroscopic measurements of adsorption isotherms, zeta potential, wettability and calorimetry [5]. These studies have been helpful for speculating mechanisms that govern the formation of the adsorbed layers on solids. Of particular interest here, a major driving force for the adsorption of long-chain surfactants has been considered to result from the self-aggregation tendency of the surfactant molecules at the solid-liquid interface. Such surface aggregation leads to the formation of two dimensional aggregates called hemi-micelles.

Hemi-micellization occurs as a result of the high affinity of the solid surfaces for the surfactant. When polymers and other additives are present, additional microstructures can form as a result of polymer-surfactant interactions with marked effect on such properties as adhesion.

Long-chain fatty acids which chemisorb on metal oxides are better lubricants than simple hydrocarbons. This is so because, possibly, the former orient themselves normal to the surface to form a hemi-micelle type structure whereas the latter absorb weakly and do not lead to organized structures. It must also be noted that with increase in temperature and pressure associated with the friction, such structures can break up or reorganize due to both the effect of such variables and the molecular motions and to possible chemical conversion of adsorption sites on the solid surfaces.

It must be abundantly clear that adsorbed layer structure in terms of molecular packing and orientation has to be considered as the primary property that controls the lubrication behavior of adsorbed films and should receive attention from researchers. Also microviscosity or fluidity on a nanoscale and its variation with temperature and pressure should be investigated along with interactive forces between the film under such conditions.

The difficulty in studying the above properties heretofore had been the unavailability of techniques for reliable *in-situ* investigation of solid-liquid interface [6]. Techniques such as infrared spectroscopy and ellipsometry involve ex-situ procedures such as freezing, drying, in vacuo, etc. and/or have necessitated the use of ideal surfaces. Clearly these methods have limited applicability for studying surfactant adsorption on particulate solids in aqueous media.

Advances in fluorescence and ESR spectroscopy have enabled the use of organic fluorescent probes for in-situ characterization of molecular environments. The utility of fluorescence methods in such studies arises from the fact that the fluorescence and ESR responses of numerous probes are highly environment dependent so that specific information may be obtained by appropriate choice of probes. Fluorescence responses that have been shown to depend on micellar environment include excitation and emission, fluorescence polarization and quenching (or sensitization). These responses have been, in turn, related to molecular properties such as polarity, viscosity, diffusion, solute partitioning and aggregation numbers.

Recently we have successfully used pyrene and dinapthylpropane (DNP) fluorescence and nitroxide ESR probes to investigate the structure of the adsorbed layer of sodium dodecylsulfate at the alumina-water interface [7]. The fluorescence fine structure of pyrene yielded information on the polarity of the microenvironment in the adsorbed layer. Intramolecular eximer formation of DNP was used to measure the microfluidity of this environment. The results indicate the presence of highly organized surfactant aggregates at the solid-liquid interface, formed by the association of hydrocarbon chains. Fluorescence decay methods enabled the determination of the size of these aggregates and their evolution as a function of surface coverage (see Fig. 1).

Three nitroxide ESR probes (5, 12, 16-DOXYL stearic acid) were used to investigate the structure of sodium dodecylsulfate interfacial layer between water and alumina formed by SDS (hemimicelles of SDS). It was found that the microfluidity within the surface aggregate varies according to the distance from the solid surface: the nitroxide sees a more ordered environment closer to the solid surface, which represents the first report of variations in flexibility (microviscosity) within a surface aggregate.

Adsorption of the above surfactants and the structure of the adsorbed layer will depend to a large extent on the nature of the solid surfaces involved. In the case of metals subjected to friction, it would therefore be important to simultaneously characterize the surface particularly with respect to its oxidation state using such techniques as ESCA. With the surface and the adsorbed layer characterized well, it would prove very useful to investigate the forces between surfaces and alterations in the forces when the surfaces are separated by only a few namometers. This can now be done easily using devices that have developed by Tabor and Israelachvili [8]. Furthermore, changes in the forces with films that have been subjected to high temperatures and pressures should also be investigated.

CONCLUDING REMARKS

Behavior of adsorbed layers on a *molecular* scale must be recognized as a critical process in tribology. There is a dire need to investigate the nanoscopic behavior of adsorbed layers using spectroscopic and various interfacial techniques that have now become available and to establish the role of various additives and environmental (e.g., temperature and pressure) factors on controlling such behavior.

REFERENCES

1. a) "Principles of Flocculation, Dispersion and Selective Flocculation," P. Somasundaran, in *Fine Particles Processing*, P. Somasundaran, editor, *2*, American Institute of Metallurgical Engineers, New York, 947-976 (1980).
   b) "Precipitation Redissolution of Sulfonates and Their Role in Adsorption Minerals," P. Somasundaran, M. S. Celik and A. Goyal, in *Surface Phenomena in Enhanced Oil Recovery*, D. O. Shah, editor, Plenum, New York, 641-647 (1981). "Surfonate Precipitation-Redissolution-Reprecipitation in Inorganic Electrolytes," M. S. Celik, E. D. Manev and P. Somasundaran, *American Institute of Chemical Engineers Symposium Series*, *78*, 86-96 (1982).
   c) R. Nagarajan, Personal Communications.
2. "The Influence of Molecular Structure on the Frictional Behavior of Lubricating Fluids," K. H. Hentschel, *Journal of Synthetic Lubrication*, *2*, 143-165 (1985).
3. "Tribology, Lubricants, and Lubrication Engineering - A Review," W. J. Bartz, *WEAR*, *49*, 1-18 (1978).
4. "Effect of Additive Molecular Structure on Friction Coefficient and Adsorption," S. Jahanmir and M. Beltzer, *Journal of Tribology*, *108*, 109-116 (1986).
5. a) "Surfactant Adsorption at the Solid-Liquid Interface - Dependence of Mechanism on Chain Length," P. Somasundaran, T. W. Healy and D. W. Fuerstenau, *Journal of Physical Chemistry*, *68*, 3562 (1964).
   b) "Adsorption of Ionic Surfactants," D. B. Hough and H. M. Rendall, in *Adsorption from Solution at the Solid-Liquid Interface*, G. D. Parfitt and C. H. Rochester, editors, Academic Press, New York, 247-319 (1983).

c) "A Study of the Interactions Between Particles and Bubbles in Surfactant Solutions," P. Somasundaran, P. Chandar and K. Chari, *Colloids and Surfaces*, *8*, 121-136 (1983).

6. "Preparation and Characterization of Clean Mineral Surfaces," P. Somasundaran and B. M. Moudgil, *Surface Contamination*, *1*, K. L. Mittal, editor, 457-475 (1979).

7. a) "Fluorescence Probing of Microfluidity of Surfactant Layers at the Solid-Liquid Interface," P. Somasundaran, N. J. Turro and P. Chandar, *Colloids and Surfaces*, *20*, 145-150 (1986).

   b) "Variation of Probe Chain Mobility in SDS Hemimicelles," P. Chandar, P. Somasundaran, K. C. Waterman and N. J. Turro, *Journal of Physical Chemistry*, *91*(1), (1987).

8. a) "The Direct Measurement of Normal and Retarded van der Waals Forces," D. Tabor and R. H. S. Winterton, *Proceedings of the Royal Society (London)*, *A312*, 435-450 (1969).

   b) "Direct Measurement of Long Range Forces Between Two Mica Surfaces in Aqueous KNO_3 Solutions," J. N. Israelachvili and G. E. Adams, *Nature*, *262*, 774-776 (1976).

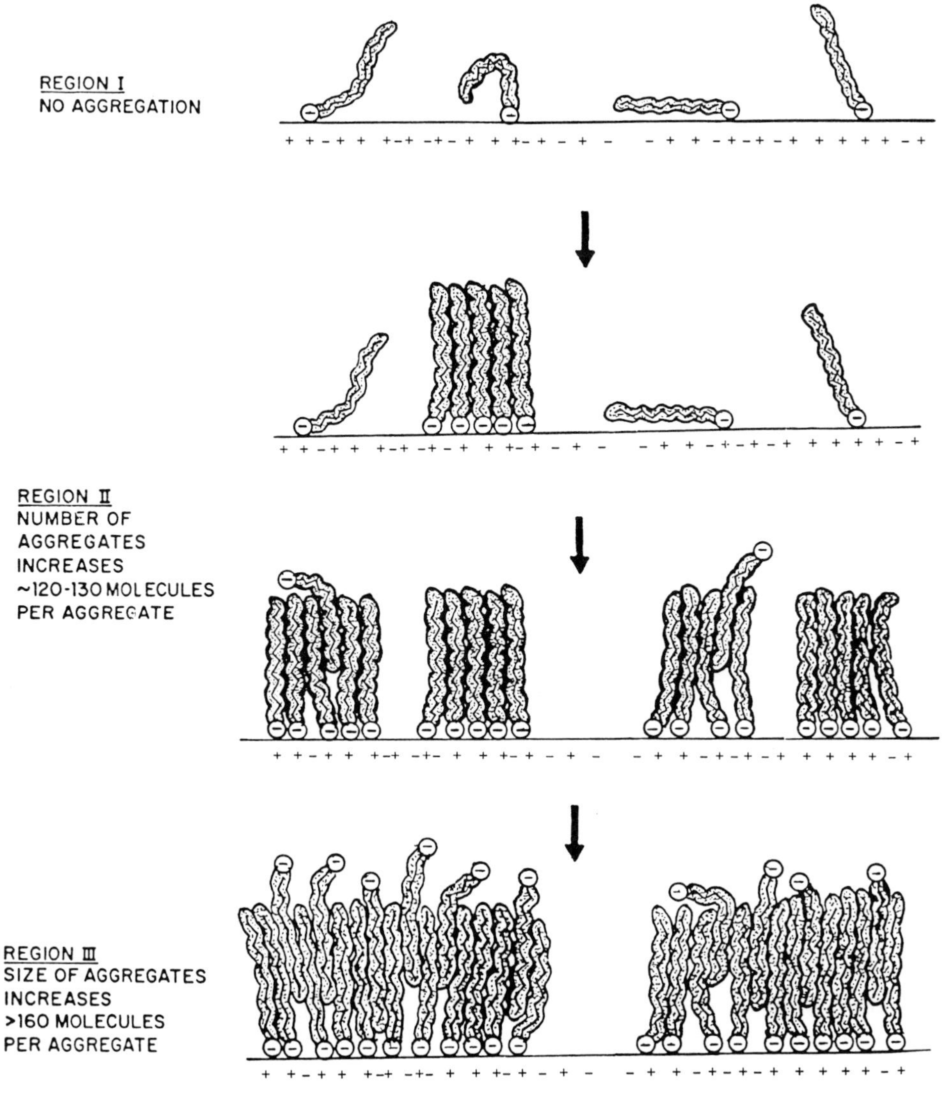

Fig. 1. Evolution of Surface Aggregates at the Solid-Liquid Interface.

4.4 A PROPOSED MODEL FOR THE DEVELOPMENT OF HIGH TEMPERATURE FLUIDS FOR LUBRICATION OF CERAMICS

David A. Dalman

The following discussion will attempt to outline a plan for overcoming at least one current impasse in the utilization of engineering ceramics in high temperature sliding contacts and illustrate the possible role of the synthetic chemist coupled with other disciplines in designing new and effective tribological systems.

Engineering ceramics are being proposed for use in high temperature applications such as the "adiabatic" engine, automotive gas turbine, roller and sliding bearings, and metal cutting tools. Attractive properties of ceramics for these uses include low density, high hardness, and high corrosion resistance over a wide temperature range. Their chief drawbacks seem to be poor reliability and uneconomical processing techniques.

When ceramics were considered for use in high temperature sliding applications, some assumed that since ceramics in general had such high values of hardness that they would show very low levels of wear (since hardness generally appears in the denominator of wear equations). It was also assumed (apparently mostly by non-chemists) that if ceramics in sliding applications did need to be lubricated, that only solid lubricants would withstand the extreme temperatures involved.

It has now been shown by several investigators that dry sliding of ceramics on themselves will not give satisfactory coefficients of friction or service life [1]. Thus, there are studies in progress on solid lubrication of ceramics, ion implantation of surfaces, composites, and cermets. Not much data is yet available but at least in the case of solid lubricants the issue of renewal on the surface poses a difficult engineering dilemma. As compared to those technologies mentioned above, the use of liquid lubricants, if they were available, would be helpful in that conventional engineering systems would still be applicable.

In the model for development of liquid lubricants for use with ceramics, three areas need to be addressed: (1) synthesis of a stable fluid basestock, (2) synthesis of additives which are both thermally stable and interact with ceramic surfaces to provide lubrication in the highly loaded system where fluid film strength is no longer sufficient, and (3) understanding of the environmental effects of both fluid basestock and surface adsorbed additives on the stress required to fracture the substrate material.

I. SYNTHESIS OF THERMAL-OXIDATIVELY STABLE BASESTOCKS

The fluid stability required to meet the needs of engine systems insulated with ceramic components has been estimated at 350-700°C at the hottest position. Since residence time for the fluid in this region is short, however, the bulk fluid does not see these temperature extremes. It is doubtful that incremental improvements in existing commercial synthetic or experimental lubricants will be able to satisfy these requirements, since an additional 100-300°C increase in stability is needed. Likewise, it is doubtful that additive packages developed for mineral oils will survive the rigorous conditions imposed by sliding ceramics at high temperatures. It is, therefore, likely that *all new chemistry will be required.*

With the above as background one might then consider the role of the synthetic organic/inorganic/organometallic chemist in designing molecules to meet the needs of this new technology.

Many organic chemists, for example, feel that we are entering into an age in which we will be able to design molecules to make them do what we want them to do. Computer programs are now available to correlate structure with properties for organic molecules and this technique is being adapted to the design of new active agricultural, pharmaceutical and polymer products. Over the next few years these quantitative structure-activity relationships (QSAR) will hopefully be extended to the correlation of lubricant properties with structure.

If the normal properties of a ceramic lubricant basestock are that it be fluid at -40°, stable at 500°C in the presence of ceramic surfaces, have high VI, and produce minimum varnish and sludge upon decomposition all at a reasonable cost for the application, then several features of such a fluid can be immediately addressed.

First is the issue of thermal-oxidative stability. Here we are dealing with a kinetic phenomenon, not some intrinsic thermodynamic property of the fluid. That is, what we have to deal with here is a question of the rate of decomposition at a given temperature. Thus we can think in terms of the loss of 1% of our fluid per 1000 hours at 500°C. Assuming that a fluid is in its highest oxidation state (a requirement in our designed molecule since oxidative pathways for decomposition usually lower stability by 80-100°C) it has been shown that a direct correlation exists between bond energies and rate of thermal decomposition [2]. Use of Arrhenius plots can relate the rate of thermal decomposition at a given temperature to bond energies. For example, construction of a molecule with no bond dissociation energies of less than 90 KCAL/mole should achieve a fluid with a rate of decomposition of 1%/1000 hours at 500°C, assuming homologous free radical dissociation as the breakdown mechanism.

It thus appears that achieving a desirable order of thermal stability in organic molecules is theoretically possible. But, it is also readily clear upon inspection of the type of bonding used to achieve this order of stability that these same molecules will be solids at room temperatures. This problem can be approached in one of two ways. One approach would be to disrupt the molecular symmetry in the structure and the other would be to use the formation of eutectic mixtures of molecules to depress freezing point. The latter approach is commonly used in commercial heat transfer fluids.

The other issue which could be raised about the basestock is its tendency to decompose into undesirable by-products. Less is known about this in relationship to molecular structure, but surely decomposition to volatile products, as occurs with some polyethers, would be preferable to formation of sludge and varnish.

Whether all of the desired features can be built into the same molecule will certainly be a challenging exercise for the organic synthesis chemist. Besides the normal carbon building blocks the synthesis chemist has at his disposal, inorganic structures such as B-C, B-N, P-N, etc. should be looked at for lubricant uses. A lot of work has been done on C-F chemistry to overcome the oxidative instability of the C-H bond and even the C-D bond [3] has been reported to decrease the rate of thermal decomposition by a factor of 10-14 fold.

II. SYNTHESIS OF LOAD-BEARING ADDITIVES

If the mechanism of action of extreme-pressure additives in mineral oil based lubricants is taken as the tribochemically induced reaction of alkyl sulfur, halogen, or phosphorus compounds with iron surfaces to form low shear stress films for the prevention of seizure, then one would want to consider the possibility of creating additives to behave in a similar manner on ceramic surfaces.

A couple of things need to be considered. First, there is no apparent chemical pathway for these conventional additives to react with ceramic surfaces such as SiC, Si_3N_4, or ZiO_2 in the same manner as the reaction with iron or other metal surfaces. Secondly, since it appears that ceramics do not have the same tendency to adhere under high loads as do metals but rather undergo some minimal plastic deformation or more likely brittle fracture, it would be well to consider whether or not the formation of a low shear stress film would increase the stress level required to fracture the material.

Again, the efforts of a synthetic organic and inorganic chemist along with the knowledge of a ceramist could be combined to design load-bearing additives for use with ceramic lubricants. Sacrificial molecules with a non-polar alkyl moiety coupled with a polar reactive end, which would be surface active and react at high contact temperatures and pressures, would be a synthesis concept for constructing a model.

III. ENVIRONMENTAL EFFECTS

Finally, one needs to consider the environmental effects of the basestock and surface adsorbed additives on the fracture toughness of the ceramic substrate. Since the toughness of current ceramic materials is marginal with respect to the desired applications, the last thing needed is the presence of chemical species on the surface which further lowers the stress intensity required to initiate and propagate cracks.

It has been well documented that water, moist air [4], aqueous solutions of surfactants [5] and polar solvents have a deleterious effect on the level of stress required to propagate cracks in brittle materials including soda-lime glass, geologic materials, quartz, and ceramics.

The mechanism of this chemomechanical effect is not precisely known but is somehow related to a lowering of the surface free energy or a stress corrosion effect of silanol bonds [6]. A better understanding of the chemical environment at the propagating crack tip would greatly help in the development of additives whose effect would be to maintain or enhance the fracture resistance of ceramic materials. The coupling of the interest and knowledge of fracture mechanics people along with that of surface scientists, tribologists and ceramists to achieve correlation of environmental effects and wear would advance this field immensely. Use of the newest techniques in surface science such as the scanning tunnelling microscope [7], which can be used at ambient conditions, could produce breakthroughs in this area and should be encouraged. Additives which promote etching and crack blunting in order to raise the stress required to propagate cracks could be added directly to a lubricant formulation or could possibly be derived as a by-product of the decomposition of the basestock.

In summary, it appears from initial theoretical considerations that the design and synthesis of a fluid lubricant for use in ceramic service is feasible. The development of fluids for this application would have the advantage of avoiding the need for significant engineering design changes in order to use ceramic materials. This technology will have to be developed concurrently with that of the ceramic materials themselves, or likely neither will succeed commercially. The challenges in this area are formidable but the knowledge gained about the interaction between new materials would seem to be worth the effort involved.

REFERENCES

1. a) "Dynamic Contact of Ceramics," K. F. DuFrane and W. A. Glaeser, presented at the Twenty-Fourth Automotive Technology Development Contractors' Coordination Meeting, Dearborn, MI, October 30, 1986.
   b) "Sliding Seals for Adiabatic Engines," S. Vinyard, et al., presented at the Twenty-Fourth Automotive Technology Development Contractors' Coordination Meeting, Dearborn, MI, October 28, 1986.
2. *Properties of Polymers, Their Estimation and Correlation with Chemical Structure*, D. W. Van Krevelen, Second Edition, Elsevier, New York (1976).
3. "The Determination of the Oxidative Stability of Several Deuterated Lubricants by an Electronic Gas Sensor," H. Ravner and H. Wohltjen, *Preprint 82-AM-SA-4*, American Society of Lubrication Engineers, 1-4 (1982).
4. a) "Observations on the Sliding Wear of Ceramics," C. S. Yust and F. J. Carignan, *American Society of Lubrication Engineers Transactions*, *28*, No. 2, 245-252 (1986).
   b) "Friction and Wear of Hot Pressed Silicon Nitride and Other Ceramics," H. Ishigaki, et al., *American Society of Lubrication Engineers Transactions*, *108*, 514-521 (1986).
5. "Surfractants," J. D. Dunning, *Chemtech, 10*, 356-359 (1980).

6. a) "The Roles of Surface Adsorption and Mobility in Geologic Systems and Materials," J. D. Dunning, *Surface Mobilities on Solid Materials*, V. T. Bink, editor, Plenum Publishing Corporation, New York, 541 (1983).

b) "Mechanisms of Subcritical Crack Growth in Glass," S. M. Wiedershorn, *Fracture Mechanics of Ceramics*, R. C. Bradt, et al., editors, 4, Plenum Press, New York, 549-580 (1978).

7. "A Crystal-Clear View of the World's Tiny Wonders," S. Dickman and O. Port, *Business Week*, 134, November 3, 1986.

4.5 SIMPLE MODEL OF METALWORKING FRICTION UNDER EXTREME PRESSURE

Frederick F. Ling and Marshall B. Peterson

INTRODUCTION

This model was first introduced in the late 1960's [1,2]. We are revisiting it because of its simplicity on the one hand and for what questions it was able to answer on the other hand. During the mid 1960's, the Materials Advisory Board of the National Research Council asked two questions regarding metalworking friction under high pressure. The prevailing understanding at the time was that friction under high pressure should have an upper limit in the shear strength of the metal being worked, as calculated from simple tensile test data [3]. The first question was why the aforementioned was often not borne out by experience.

Moreover, the prevailing understanding at the time was that shear strength of metals, at least those being commonly worked, was a constant mechanical property under isothermal condition. According to this understanding the coefficient of friction, which is bounded by the ratio of the shear strength and the hydrostatic pressure, should approach zero as pressure increases indefinitely. The second question raised was why this was not borne out by experience. Fig. 1 shows a typical coefficient of friction vs. pressure curve for unlubricated sliding; that is, the coefficient of friction reaches an asymptote which is non-zero.

EXPERIMENTAL OBSERVATIONS

A number of experiments were carried out by one of us (MBP) while the theoretical model was being developed. In order to accentuate the high pressure and what we called extreme pressure phenomena, we decided to look at the coefficient of friction as a function of the inverse of the pressure. Fig. 2 shows such a transformation of the data shown in Fig. 1. In addition, Fig. 2 delineates the inverse pressure scale into four regimes as indicated. Regime I is the first constant friction regime where Amontons' First Law holds; it is characterized by low average interface pressure. Regime IV, which we called the extreme pressure regime, is the second constant friction regime. Regime II is the usual metalworking regime. Historically, the slope of the f vs. p^{-1} curve has been taken to be the shear strength of the material being worked. By this method of representation it is clear whether the slope of a given material matches the shear strength of the material. In gener al, the slopes do not match the shear strengths of the materials studied. Regime II is the transition between Regimes III and I. Fig. 3 shows selected data on coefficient of friction f vs. die surface roughness [4].

SIMPLE MODEL OF METALWORKING FRICTION

With the aforementioned experimental observations, we proceeded to the simple model, which follows, so as to answer the two questions: (1) Why the slope of the f vs. p^{-1} curve in Regime III is not equal to the shear strength of the material being worked in general? (2) Why f has non-zero value(s) in Regime IV? Fig. 4 shows one of the experimental schemes which simulates the tool and workpiece together with an enlargement of the interface. This enlargement shows a two-dimensional model of a typical repeating segment of metal being worked which is trapped between two peaks on the die surface. This segment has been further idealized and is shown in Fig. 5. When the material is assumed to be rigid plastic having a single mechanical property in the shear strength, the method of slip-lines can be applied to solving the coefficient of friction $f = F_i/L_i$. F_i and L_i are shown in Fig. 5. For this problem we established the solution [2,5]:

$$f = B \frac{k}{p} + \frac{s_i}{p} \quad , \tag{1}$$

where

$$B = \frac{1}{2}\left[1 + \sqrt{1 - (s_i/k)^2} + 4\beta - \frac{\pi}{2} - \cos^{-1}(-s_i/k) \right]\cot \beta \quad ,$$

k is the shear strength of the metal being worked, s_i is a surface effect index which can be a measure of adhesion or lubricity as the case may be, and β is the semi-angle of the model wedge shown in Fig. 5. Equation (1) shows why the slope of f vs. p^{-1} curve is, in general, not equal to k. Rather there is a geometric factor with surface effect involved and a surface factor as discussed, among others. Moreover, we have shown [6] that k is pressure sensitive beyond a certain threshold. For simplicity, beyond a certain threshold, we write $k = k_o + \alpha p$, where α is the coefficient of pressure dependence. Now, when this is introduced into Equation (1) and for p very large, we have

$$f \simeq B\alpha \quad , \tag{2}$$

which is the friction at extreme pressure.

Using Equation 1, Fig. 6 is constructed. That is, the dependence on the semi-wedge angle β of dimensionless friction. Referring to Fig. 3 and Fig. 6, though, the simple model does capture the effect of surface roughness in a semi-quantitative way. Of course, surface roughness plays an important part in B which in turn governs, to a significant degree, the slope of the f vs. p^{-1}, see Equation 1, and f at extreme pressure, see Equation 2.

Since this model was published two decades ago, there have been many papers which used slip line theory to predict friction and even wear (e.g., [7]) in what are Regimes I and II according to Fig. 2.

REFERENCES

1. "Friction and Lubrication in Metalworking Processes," F. F. Ling and M. B. Peterson, *Proceedings of the International Conference on Metalworking Technology*, American Society of Tool and Manufacturing Engineers, 1181-1192 (1967).
2. "Mechanics of Sliding Surfaces," F. F. Ling, *Metals Engineering Quarterly*, *10*, 1-4 (1967).
3. *Mechanics of Plastic Deformation in Metal Processing*, E. G. Thompson, C. T. Yang and S. Kabayashi, The MacMillan Company (1965).
4. "Frictional Behavior in Metalworking Processes," M. B. Peterson and F. F. Ling, *Journal of Lubrication Technology, Transactions of the American Society of Mechanical Engineers*, *92*, 535-542 (1970).
5. *Surface Mechanics*, F. F. Ling, Wiley Interscience, New York (1973).
6. "Pressure Dependence of Shear Strength and Sliding Friction," F. F. Ling and J. D. Vasilakis, *Journal of Applied Mechanics*, *37*, 192-195 (1970).
7. "An Explanation of the Different Regimes of Friction and Wear Using Asperity Deformation Models," J. M. Challen and P. L. B. Oxley, *WEAR*, *53*, 229-243 (1979).

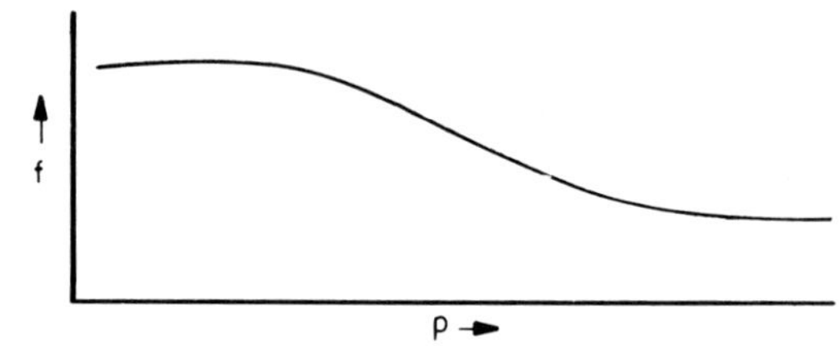

Fig. 1. Typical Coefficient of Friction f vs. Pressure p Curve Covering the Metalworking Range.

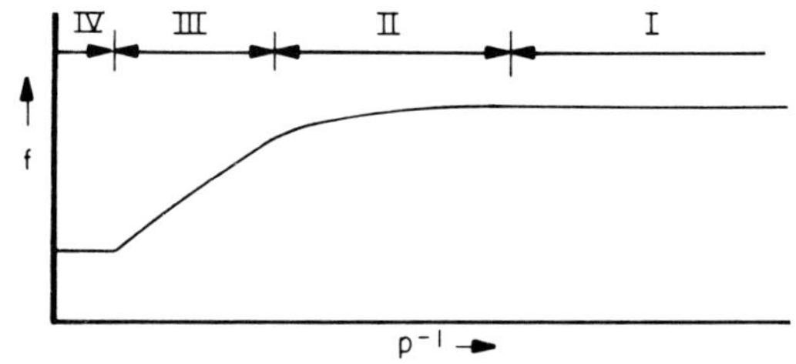

Fig. 2. The Data in Fig. 1 with the Abscissa Changed to p^{-1}.

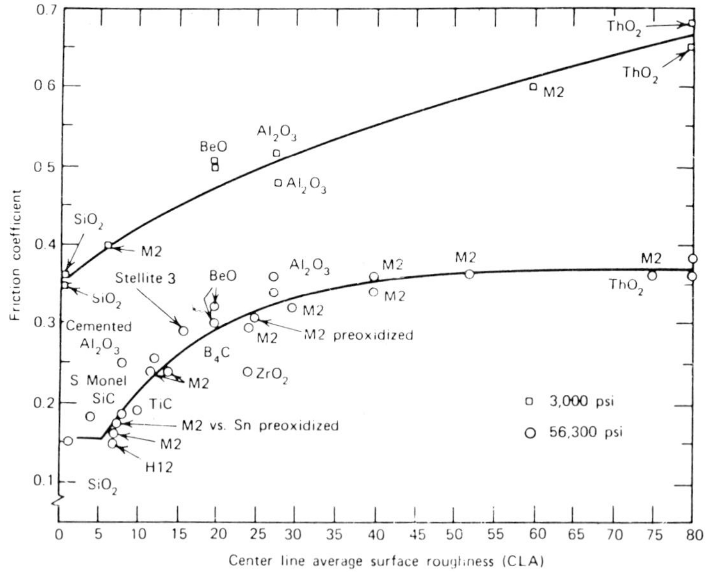

Fig. 3. Selected Data on Friction Coefficient vs. Die Surface Roughness.

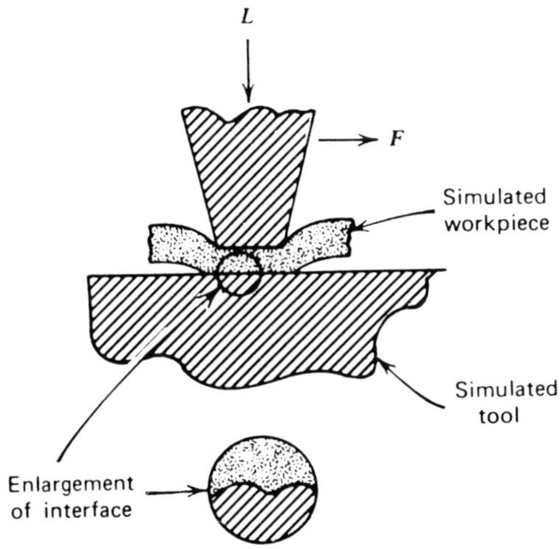

Fig. 4. Diagram of the Experimental Scheme.

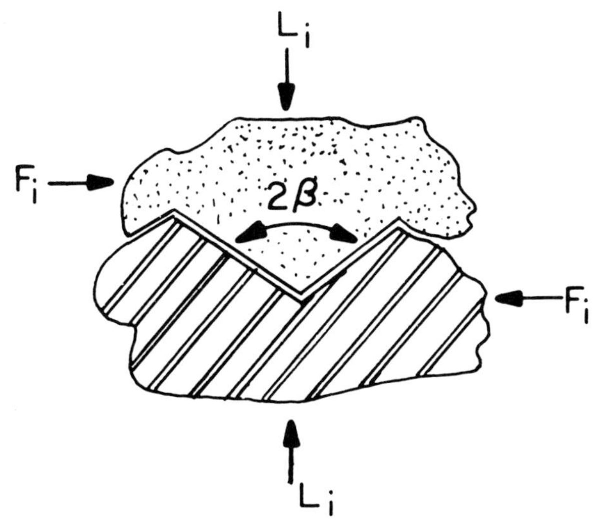

Fig. 5. Idealized Model of Worked Material and Two Peaks on the Die Surface.

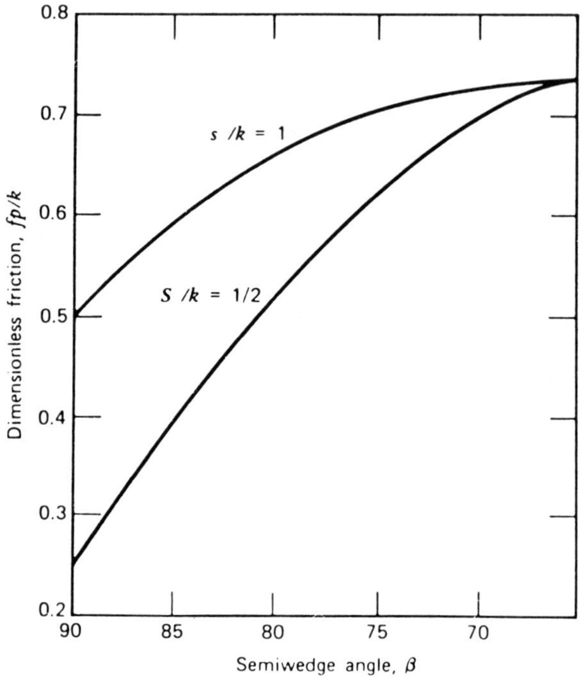

Fig. 6. Dimensionless Friction vs. Semiwedge Angle β.

4.6 USE OF CUTTING FORCE IN DISCIPLINING RELATIONS BETWEEN ABRASIVE WEAR AND MECHANICAL PROPERTIES

Jorn Larsen-Basse

INTRODUCTION

One of the simplest wear processes is two-body abrasion of ductile metals by hard abrasives, as in the case when a metal sample slides over an abrasive paper. In this situation, metal removal is by surface deformation and micromachining, and it would appear relatively simple to model the process from machining theory and mechanical behavior at large strains and strain rates. This has not yet happened, however, in any quantitative sense, although reasonably good quantitative explanations can be developed for most types of observed behavior.

In most abrasion tests the conditions of load, abrasive type and size, etc. are fixed and the material is varied. The wear resistance is plotted vs. hardness and a reasonable fit is generally obtained. This is consistent with a simple model of abrasion where sharp microtools indent the surface to an extent which is determined by its hardness, and where these microtools cut neat grooves during movement of the sample.

Real abrasives are not perfectly hard nor perfectly sharp. They will deform elastically when contacting a metal surface which has a hardness approaching their own, and the principle of geometric similarity is therefore not maintained. This makes direct comparison of different materials difficult both when removal is by groove formation and when it is by brittle cracking. In this note, only the situation where very hard abrasives (SiC) are abrading relatively much softer metals (hardened and tempered carbon steel) is considered. In this case, testing under constant load will give results for conditions where the indentations are not geometrically similar between the various levels of sample hardness, as illustrated in Fig. 1. Since the real abrasives have somewhat rounded corners and since groove depth is only a small fraction of the abrasive diameter the indenter should be represented as a rounded shape. In that case different depths of indentation caused by loading at constant load against samples of different hardness will result in geometrically dissimilar indentations. Then the surface strain from the indentations themselves will be different as it is related to the ratio between depth and radius of curvature; and the conditions at the "tool" front will be different with different average attack angles, relative amounts of material pushed up as chips vs. the amounts pushed to the groove sides, etc.

For these reasons it would appear to be more useful to compare different materials based on the conditions which will produce grooves of equal depth. However, since wear rate generally is found to be proportional to load and groove depth may be a function of load, it is possible that wear rate itself has insufficient sensitivity in this regard. One compound property which may be more useful, although it is rarely measured, is the cutting force, since its magnitude should directly relate to the mechanics of material deformation in and around the groove.

In this note cutting force data are evaluated in some detail to illustrate the usefulness of this property in more detailed abrasive studies. The data were obtained under conventional constant load conditions.

EXPERIMENTAL

Cylindrical specimens 3.7 mm in diameter were abraded dry under 1 kg load against 180 grit (83 μm) SiC abrasive paper at a sliding velocity of 37 mm/s. Samples were AISI 1045 steel hardened and tempered to a full range of hardnesses.

RESULTS

The wear resistance-hardness plot shown in Fig. 2 is similar to data presented by others also. The behavior of these materials does not follow the simple wear resistance-hardness proportionality expected from simple theory and this has spawned a number of proposed explanations. The cutting work to remove unit volume of material follows a similar relationship while the cutting force appears to be more directly related to sample hardness.

SEM observation of abraded surfaces shows that each sample hardness results in an essentially constant groove width, greater for the softer materials. The hardest samples have essentially no material pushed to the groove sides while the amount increases for the softer samples which makes it more difficult to determine groove size.

DISCUSSION

Since the hardest sample (790 H_V) has the most clean-cut grooves it has been used as a basis for the calculations below. The number of contacting abrasive grits, N, is obtained from

$$N \times \frac{\pi}{4} \times w^2 \times H = P \qquad (1)$$

where H is the hardness, w is the groove width (measured at 7.1 μm), and P is the applied load. The result is N = 32 which would appear to be reasonable, considering the load and the speciment diameter. The groove depth is estimated by assuming a triangular groove cross-section and complete removal of the groove material as wear

$$N \times \frac{1}{2} wd \times L = WR \qquad (2)$$

where d is the groove depth, L is the reference sliding distance (1 m) and WR is the volumetric wear per m. The result is d = 1.2 μm, which appears consistent with SEM observations.

For the more ductile samples the plowing component of the groove formation tends to distort the groove width measurements. For these materials a different approach was used.

Since differences in groove depth between the softest and the hardest material is less than 1.5 μm vs. a grit diameter of 83 μm, it is reasonable to assume that N remains constant. Then w can be calculated from Equation (1). If d is estimated from assumed geometric similarity with the martensitic sample then the horizontal stress component, S can be calculated from

$$F = N \times \frac{1}{2} dw \times S \tag{3}$$

where F is the measured cutting force.

Values of w range from 7 to 15 μm and are plotted in Fig. 3, together with some measured values. Values of d range from 1.2 to 2.6 μm. The ratio S/H shows a monotonic decrease with hardness from 6.3 for the softest material to 4.7 for the hardest, as also shown in Fig. 3. Considering the large strains and strain rates in abrasion these numbers appear to be in the expected general range.

The estimated fraction removed, f, can now be obtained from a modification of Equation (2)

$$N \times \frac{1}{2} wd \times L \times f = WR \tag{4}$$

This value ranges from one (baseline) for the martensitic sample to 0.4 for the fully annealed sample. The data are plotted in Fig. 3.

While a number of simplifying assumptions were made in developing these calculations the results clearly show that the abrasion resistance of these materials is determined by a combination of flow stress and strain to fracture of that portion of the material which is being pushed to the groove sides by plowing action.

CONCLUSIONS

The usefulness of cutting force measurements in evaluating abrasive wear mechanisms has been demonstrated. It has been shown that abrasion resistance of hardened and tempered steels is closely determined by flow stress and strain to fracture in the plowing process.

ACKNOWLEDGEMENTS

The wear data were obtained by Young-Ho Park at the University of Hawaii.

REFERENCES

1. "Resistance of Metals to Wear by Abrasion, as Related to Hardness," M. M. Kruschov, *Proceedings of the Conference on Lubrication Wear, London*, Institution of Mechanical Engineers, London, 655-659 (1957).

2. "The Abrasion Resistance of Some Hardened and Tempered Carbon Steels," J. Larsen-Basse, *Transactions of the American Institute of Metallurgical Engineers*, *236*, 1461-1466 (1966).
3. "Influence of Structure on the Abrasion Resistance of a 1040 Steel," J. Larsen-Basse and K. G. Mathew, *WEAR*, *14*, 199-206 (1969).
4. "Abrasive Wear Mechanisms in Steels," M. J. Murray, P. J. Mutton and J. D. Watson, *Wear of Materials 1979*, K. C. Ludema et al., editors, American Society of Mechanical Engineers, New York, 257-265 (1979).

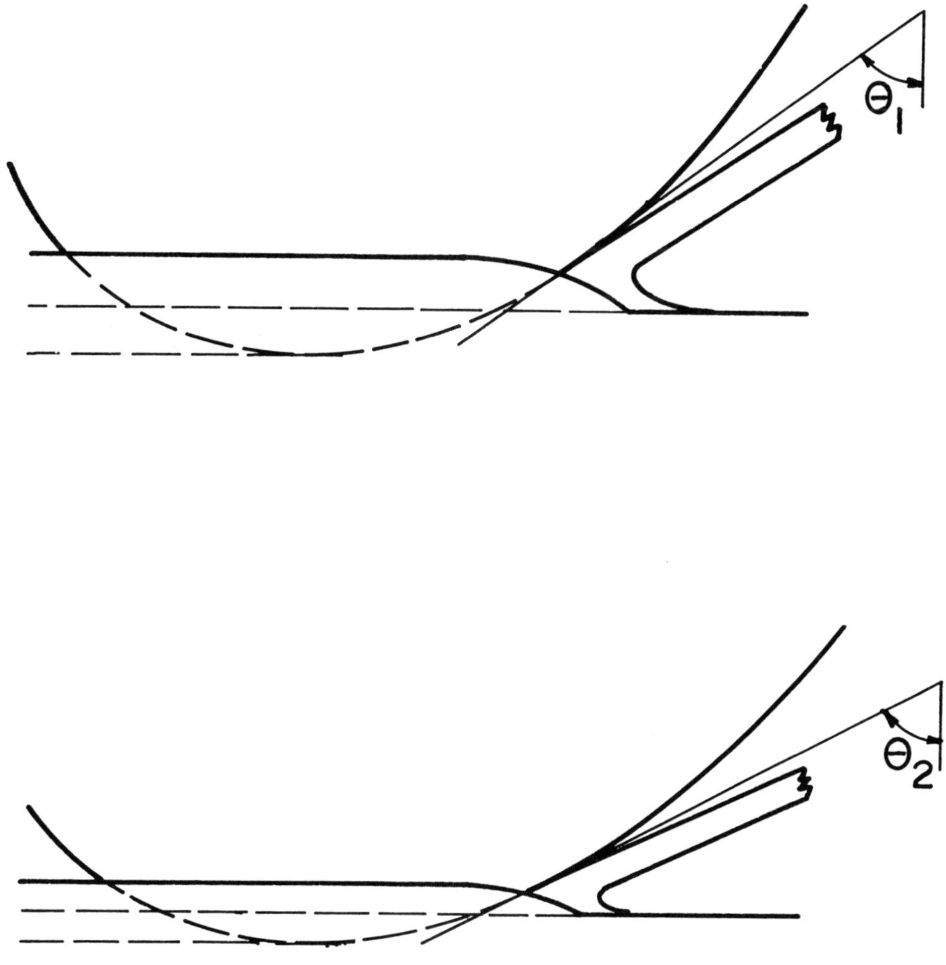

Fig. 1. Sketch showing the change in groove formation conditions with depth of penetration of a rounded abrasive into the metal surface.

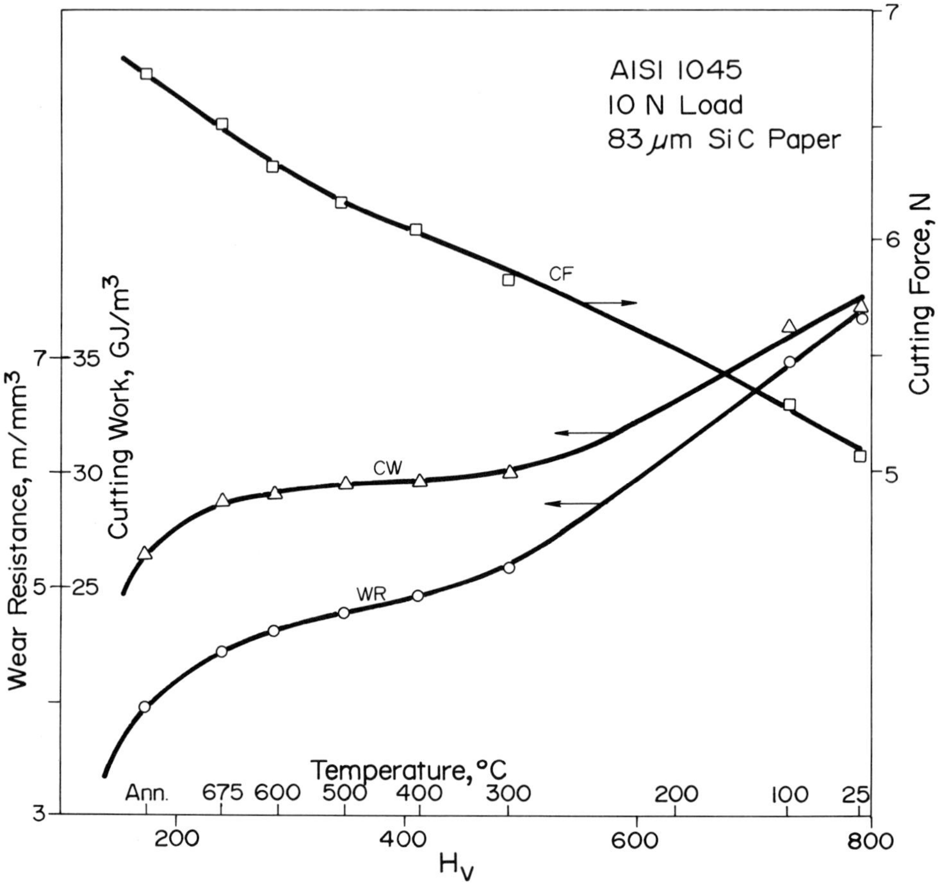

Fig. 2. Wear resistance (WR), cutting work (CW), and cutting force (CF) as functions of the sample bulk hardness for hardened and tempered AISI 1045 steel.

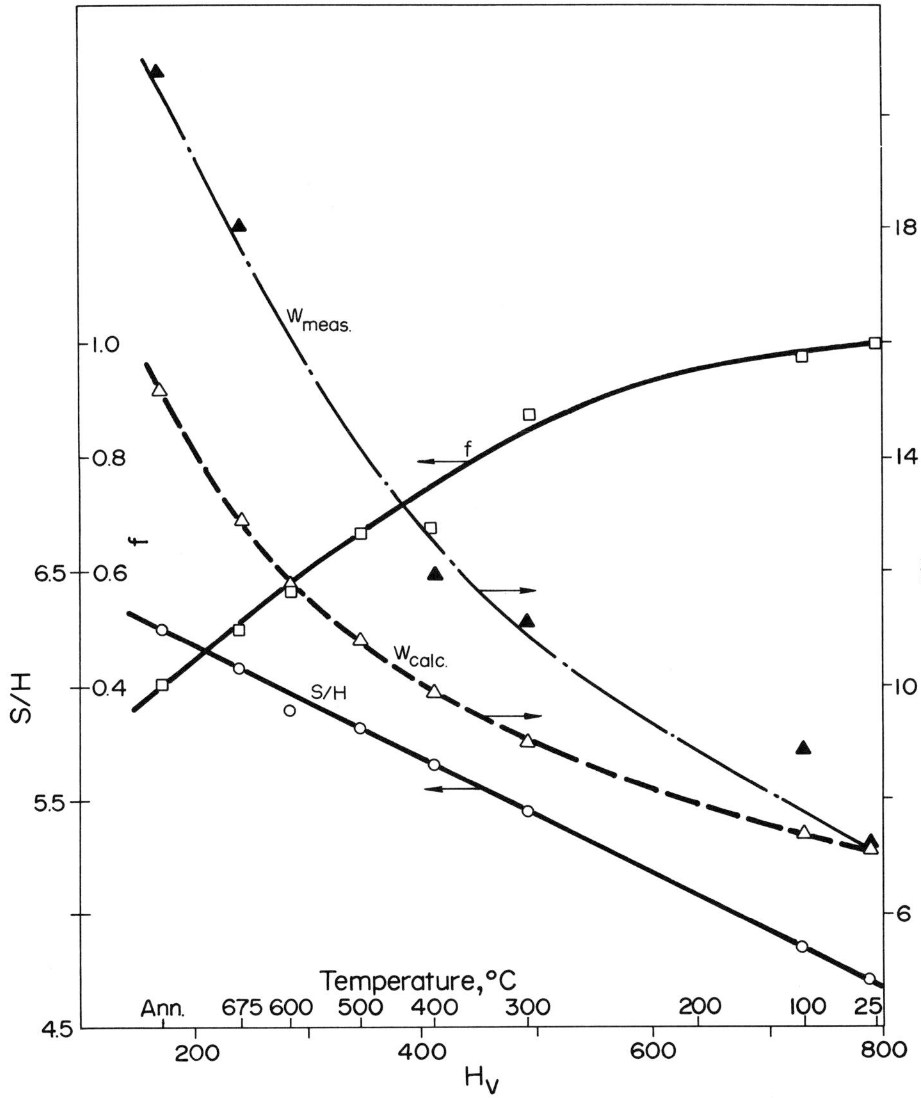

Fig. 3. Plotted vs. sample hardness are calculated (w_{calc}) and measured (w_{meas}) values of groove width; calculated horizontal stress component, S, divided by sample hardness; and calculated fraction, f, of groove cross section which is lost due to abrasion.

4.7 LARGE PLASTIC DEFORMATION IN SLIDING FRICTION AND WEAR

Francis E. Kennedy, Jr.

INTRODUCTION

It is well established that sliding friction involves large amounts of near-surface plastic deformation [1]. The deformation is caused by frictional tractions at the contact interface and is an integral part of the friction process, at least for ductile materials [2]. Microscopic examination of the highly deformed near-surface regions has shown that the plastic deformation is also responsible for most sliding wear through such mechanisms as void formation, crack nucleation and propagation, extrusion and other chip or particle formation processes [3,4]. It is therefore imperative that any models of friction and wear processes account for large near-surface plasticity effects.

Quantitative studies of plastic deformation in sliding contacts have been rare, with the most published data relying on microscopic observations of worn metallic surfaces. These observations have enabled residual strains to be measured after many sliding events had occurred on the contact surface [5] or after a single pass of a slider over the surface [6]. Those studies have shown that a region of very severe plastic deformation develops very quickly on and near the sliding surface of ductile metals, with residual plastic strains of the order 10 or greater being found even after a single sliding event [6]. Recent and anticipated improvements in microscopy techniques and surface analytical tools should allow even more knowledge to be gained about near surface plastic deformation, the effects of microstructural differences on such deformation, and the relationship between plastic deformation and various wear mechanisms. What is needed are mechanical models that can enable the observed plastic deformation, along with factors affecting it, to be analyzed. Such models would improve our capability to predict trends in friction and wear behavior and would help us approach our eventual goal: quantitative prediction of friction coefficients and wear rates.

THE GOALS OF A PLASTICITY MODEL

- A better understanding of factors affecting near-surface plastic deformation in sliding contacts.
- Information on conditions leading to crack nucleation in the heavily deformed region.
- Understanding of conditions leading to crack propagation and particle formation by low cycle fatigue, ductile fracture, extrusion, delamination, chip formation, or other mechanisms resulting from plastic deformation.
- Determination of the mechanism for dissipation of frictional energy through plastic deformation.

- The eventual goal is a model to aid in the prediction of friction coefficients and wear rates.

WHAT IS REQUIRED FOR THE DEVELOPMENT OF A PLASTICITY MODEL

- A good understanding of the mechanical behavior of materials under high strain, high strain rate conditions. Some of this information is being developed by people involved in metal forming analysis, an application with plastic strains and strain rates approaching those found in sliding contacts, but much more needs to be done in quantifying that information and making it readily accessible.
- Ability to apply realistic boundary conditions. The actual surface tractions at the sliding interface are not known and, in fact, there is little quantitative information about real areas of contact in sliding junctions.
- Ability to analyze very small regions. There are many inhomogeneities in the highly-deformed regions, e.g. dislocation cell boundaries, voids and micro-cracks, which may limit the size of domains having a given set of material properties.

SOME POSSIBLE APPROACHES FOR ANALYTICAL PLASTICITY MODELS

1. Slip Line Theory [7]
2. Upper-Bound Methods [8]
3. Finite Element Methods
    A. Viscoplasticity Models [9]
    B. Updated Lagrangian Formulations [10]

SOME SUCCESSES SO FAR

Using a slip-line field theory, Challan and Oxley [7,11] have developed a model for the sliding of a hard asperity over a softer surface. Using the model they quantified the contributions of adhesion and plowing to sliding friction and were able to predict trends friction coefficient as a function of surface roughness and adhesion for some cases [7]. Using the assumption that low cycle fatigue is the primary wear mechanism, they were also able to predict observed trends in wear coefficient [11].

Also using slip-line theory, Olver [12] developed an extrusion model for wear particle formation. The model agreed with observed behavior for a variety of cases in which a subsurface crack might have been present.

Wanheim [14] also used slip-line theory to model friction and plastic deformation beneath and ahead of a sliding hard wedge-shaped asperity. That model was used successfully to predict trends in friction as a function of asperity angle, normal pressure and yield strength.

An analytical model of friction during the steady state sliding of metals was developed by Avitzur, based on an upper-bound theorem of plasticity [8]. The model focused on energy consumed by the plastic deformation of material beneath the moving slider and in the bulge or ridge which often forms just ahead of the slider. Trends in friction predicted by the model agreed well with both experimental data and results of the slip line models of Wanheim and Challen [8].

Using a viscoplastic finite element model, this author was able to predict near-surface plastic deformation with some accuracy for cases involving single-pass sliding over fully-dense and porous materials [9,13]. A viscoplastic formulation is valid in these sliding contact cases because the near-surface plastic strains are much larger than any elastic strains which occurred during the deformation and because the plastic strain rates are so large that it is reasonable to think of velocities of the deforming solid instead of displacements. Finite element programs have been written, based on viscoplastic constitutive equations, to enable determination of velocities, strain rates, and stresses in a stationary deformable solid in contact with a rigid slider [9,13].

A typical result is shown in Fig. 1 for the case of a copper specimen in contact with a 40μm wide rigid moving slider. Each arrow in the figure represents a scaled velocity vector for a different point within the copper solid. It can be noted that the material in the top 20 or 30μm exhibits very large shear rates, with almost all flow of that material being in the sliding direction. There is a small vertical component of flow ahead of the slider and that would tend to cause a very slight bulge to form there. If the slider happened to have a sharp leading edge, as with a cutting tool or a sharp abrasive grit, the bulge could become a chip. if, on the other hand, the slider had a rounded leading edge, the bulge would be smeared out, producing even more plastic deformation. By integrating the velocities along flow lines one can obtain the total displacement of a point in the solid for a single pass of the slider over the specimen surface. Calculations of this type were compared with experimental measurements of plastic deformation in single pass sliding tests [9]. It was found that good agreement between predicted and measured deformations can be achieved as long as it is assumed that the contact surface of the solid moves at an average velocity substantially less than that of the rigid slider in contact with it [9]. The viscoplastic model has been used as an aid in the development of improved abradable materials for gas path seal applications [13].

In a recent paper, Ohmae [10] analyzed the large plastic deformation for sliding wear situations using an updated Lagrangian finite element approach. He produced evidence that void formation under highly plastified conditions can lead to crack nucleation and showed that propagation of the cracks can lead to wear particle formation. Ohmae used a zooming procedure that enabled him to analyze deformation in regions of dislocation cell size [10]. The upgraded Lagrangian formulation enables the modeling of much larger deformations than can be handled by the usual small displacement models, but it may require substantial computation time to reach equivalent plastic strains greater than 100%, as are observed in many sliding contacts.

SOME UNRESOLVED QUESTIONS AND CHALLENGES

Despite the progress that has been made in the development of large displacement, high rate plasticity models for sliding contact problems, a number of questions remain unanswered and difficult challenges remain.

What are the actual boundary conditions in a sliding contact? How much adhesion between the contacting surfaces really occurs? How does deformation cause changes in the real area of contact (junction growth)?

Better material property data are needed for ductile materials at large strains and high strain rates similar to those found in the highly deformed regions near sliding contacts.

Better ability is needed to analyze and predict crack propagation in conditions of large plastic strains and large strain rates. Current fracture mechanics procedures are inadequate.

More experimental measurements (quantitative) of near-surface plastic deformation are needed.

A closer tie between near-surface plastic deformation and friction must be developed.

Until these questions and challenges are resulved, the use of plasticity models in the prediction of friction coefficients and wear rates will remain a desirable, but unachieved, goal.

REFERENCES

1. *Friction and Wear*, I. V. Kragelskii, Butterworths, London (1965).
2. "Plastic Deformation and Sliding Friction of Metals," D. A. Rigney and J. P. Hirth, *WEAR, 53*, 345-370 (1979).
3. "The Delamination Theory of Wear," N. P. Suh, *WEAR, 25*, 111-124 (1973).
4. "Plastic Deformation Below Worn Surfaces," M. A. Moore and R. M. Douthwaite, *Met. Transactions A, 7A*, 1833-1839 (1976).
5. "Quantitative Determination of Deformation by Sliding Wear," J. H. Dantzenberg and J. H. Zaat, *WEAR, 23*, 9-19 (1973).
6. "Single-Pass Rub Phenomena - Analysis and Experiment," F. E. Kennedy, *Journal of Lubrication Technology, Transactions of the American Society of Mechanical Engineers, 104*, 582-588 (1982).
7. "An Explanation of the Different Regimes of Friction and Wear Using Asperity Deformation Models," J. M. Challen and P. L. B. Oxley, *WEAR, 53*, 229-243 (1979).
8. "A Friction Model Based on the Upper-Bound Approach to the Ridge and Sublayer Deformations," B. Avitzur, C. K. Huang and Y. D. Zhu, *WEAR, 95*, 59-77 (1984).
9. "Determination of Near-Surface Plastic Deformation in Sliding Contacts," F. E. Kennedy and L. P. Grotelueschen, *Journal of Applied Mechanics, Transactions of the American Society of Mechanical Engineers, 51*, 687-689 (1984).

10. "Analysis of the Large Plastic Deformation Involved in Wear Processes, Using the Finite Element Method With an Updated Lagrangian Formulation," N. Ohmae, *ASME Paper 86-Trib-25* (1986).
11. "Prediction of Archard's Wear Coefficient for Metallic Sliding Friction Assuming a Low Cycle Fatigue Wear Mechanism," J. M. Challen, P. L. B. Oxley and B. S. Hockenbull, *WEAR, 111*, 275-288 (1986).
12. "Extrusion - A New Wear Mechanism," A. V. Olver, in *Mechanisms and Surface Distress,* Butterworths, London, 74-80 (1985).
13. "Plasticity Analysis of Wear Phenomena, as an Aid in the Development of Abradable Material," F. E. Kennedy, K. E. Hauck and L. P. Grotelueschen, in *Mechanisms and Surface Distress*, Butterworths, London, 67-73 (1985).
14. "A Model for Friction in Metal Forming Processes," T. Wanheim and N. Bay, *Ann. of CIRP*, *27*, 189-194 (1978).

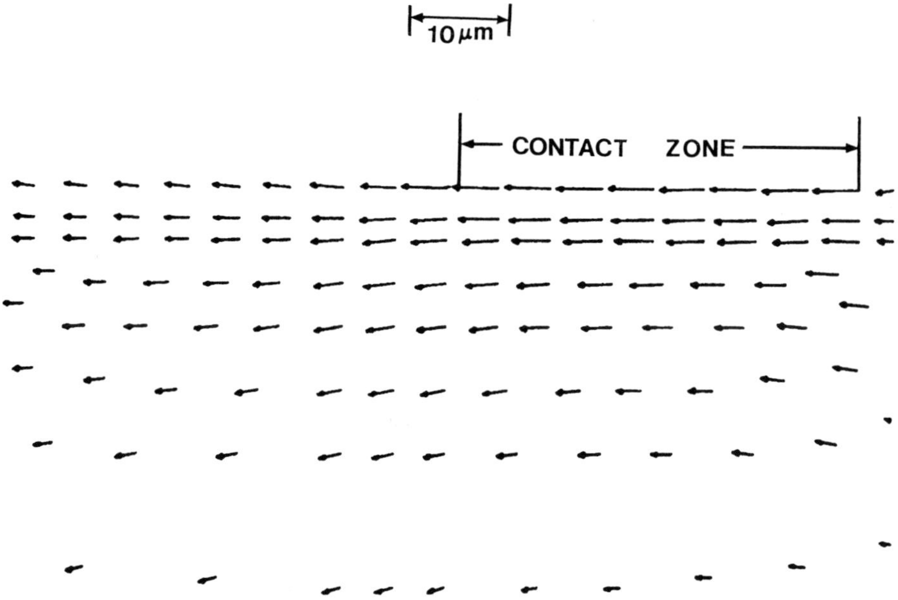

Fig. 1. Predicted Velocity Distribution in a Copper Specimen in Contact with a Rigid Slider Moving to the Left at a Velocity of 2 m/s.

4.8 FRICTION WITH SOLID LUBRICANT FILMS

Marshall B. Peterson and Mike Kanakia

INTRODUCTION

A number of models have been proposed to describe solid film lubrication [1]. One of these, and the most traditional, suggests simple shear or continuous plastic deformation of an oriented film on the surface. This hypothesis is evaluated in this brief.

ANALYSIS AND DISCUSSION

Those who have measured the shear strength of compounds at high pressures [2,3,4] have found that its behavior can be described as follows:

$$S = S_o + \alpha P$$

where S = shear strength, S_o = shear strength/low pressure, P = pressure, α = constant. Thus, the friction coefficient becomes:

$$f = \frac{S}{P} = \frac{S_o}{P} + \alpha \qquad (1)$$

If we use the shear strength as measured by Tyler and Ku [5] of 1500 psi and α from Bridgeman as 0.02, then this equation becomes:

$$f = \frac{1500}{P} = 0.02$$

This equation is plotted as a dotted line ① in Fig. 1. Also shown in Fig. 1 are the experimental data from a large number of investigations. It can be seen that the equation holds well at high pressures but does not apply at low pressures, below P*. The question then becomes what controls the friction below P*.

Above P*, the contact area represents the geometric contact area because of the high pressure. As the pressure is lowered, the friction coefficient would increase excessively if this complete contact area is maintained. Thus, it seems logical to conclude that at P* the area is reduced so friction remains low. If, according to the adhesion theory of friction, we assume that friction is controlled by the hardness then we may recalculate the low pressure friction:

$$f = \frac{S_o}{H_f} + \alpha = \frac{1550}{45440} + .02 = 0.05$$

Here we have used the value of the hardness of the oriented layer of MoS_2 [6]. If the hardness of the steel substrate is used (530 kg/mm^2) or the bulk hardness of MoS_2 (10 kg/mm^2) is measured, then quite different values are obtained.

However, it should be noted that large changes in hardness do not greatly change the friction coefficient.

These data suggest that at low pressures the adhesion theory of friction applies to thin films.

The area of contact is the "real area of contact" as controlled by the hardness of the film. As the load is increased, the area increases proportional to the load, so friction remains constant. Eventually, the real area approaches the geometric contact area and remains constant thereafter. Friction then decreases according to Equation (2).

The higher values of friction around 0.08 were obtained with pellets. If we use the bulk hardness of MoS_2, then Equation (3) becomes:

$$f = \frac{S_o}{H_f} + \alpha = \frac{1500}{25560} + .02 = 0.079$$

A friction value of 0.079 seems to fit the experimental data quite well.

This approach seems to be confirmed by the results of Barry and Binkelman [7] as shown in Fig. 2. It can be seen that the friction of MoS_2 films is independent of substrate hardness unless the substrate hardness is less than the MoS_2 film of 32 kg/mm^2.

The hypothesis appears to apply to MoS_2. More data is required to determine its applicability to other compounds.

REFERENCES

1. "Models of Solid Lubrication. Fuundamentals of High Temperature Friction and Wear with Emphasis on Solid Lubrication for Heat Engines," L. K. Ives and M. B. Peterson, *Models of Solid Lubricant Mechanisms*, F. F. Ling, editor, Industrial Tribology Institute, 43-83 (1985).
2. "Shearing Phenomena at High Pressures Particularly in Inorganic Compounds," P. W. Bridgeman, *Proceedings of the American Academy of Arts and Sciences, 71*, 387-460 (1936).
3. "The Failure of Amonton's Law in the Friction of Polymer Films Under High Pressure," L. C. Towle, *Proceedings of First International Conference on Solid Lubrication*, American Society of Lubrication Engineers, 202-215 (1971).
4. "Friction Under High Normal Pressure," R. Lauterbach, F. Lira and E. G. Thomsen, *WEAR, 10*, 469-482 (1967).
5. "Fundamental Investigation of Molybdenum Disulfide as a Solid Lubricant," J. C. Tyler and P. M. Ku, NTIS AD818 439/2EST (1967).
6. "Structure of Molybdenum Disulfide Lubricant Film," A. J. Brudney and A. F. Karmadonov, *WEAR, 33*, 243-249 (1975).
7. "MoS_2 Lubrication of Various Metals," H. F. Barry and J. P. Binkelman, *Lubrication Engineering, 22*, 139-145 (1966).
8. "The Friction Properties of Various Lubricants at High Pressures, J. Boyd and B. Robertson, *Transactions of the American Society of Mechanical Engineers, 67*, 51-55 (1945).
9. "Mechano-Chemical Factors in MoS_2 Lubrication," G. Salomon, A. W. J. deGee and J. H. Zaat, *WEAR, 7*, 87-101 (1964).
10. "The Effects of Load on the Frictional Properties of Molybdenum Disulfide," S. A. Karpe, *American Society of Lubrication Engineers Transactions, 8*, 164-178 (1965).

11. "Friction and Cleavage of Lamellar Solids in Ultra High Vacuum," D. G. Flom, A. J. Haltner and C. A. Gaulin, *American Society of Lubrication Engineers Transactions*, *8*, 133-145 (1965).
12. "Effect of Substrate Surface Finish on the Lubrication and Failure Mechanisms of Molybdenum Disulfide Films," R. L. Fusaro, *American Society of Lubrication Engineers Transactions*, *25*, 141-156 (1981).
13. "Friction of Possible Solid Lubricants with Various Crystal Structures," M. B. Peterson and R. L. Johnson, NACA TN 333L (1954).
14. "Investigation of Friction and Shear in Hot Metal Deformation," M. B. Peterson, H. S. Cheng, F. F. Ling and J. D. Vasilakis, MTI 68TR19 (1968).
15. "Performance of Conventionally Sputtered MoS_2 Versus Co-sputtered MoS_2 and Ni," B. C. Stupp, *Proceedings of Third International Conference on Solid Lubrication*, American Society of Lubrication Engineers, 217-222 (1984).
16. "Lubrication with Sputtered MoS_2 Films," T. Spalvins, *Proceedings of First International Conference on Solid Lubrication*, American Society of Lubrication Engineers, 360-370 (1971).
17. "The Effect of Metallic Substrates on MoS_2 Film Formation," J. V. Reid and J. A. Schey, *Proceedings of Second International Conference on Solid Lubrication*, American Society of Lubrication Engineers, 22-29 (1978).
18. "Intercalated Dichalcogenide Solid Lubricants," W. E. Jamison, *Proceedings of Third International Conference on Solid Lubrication*, American Society of Lubrication Engineers, 73-87 (1984).

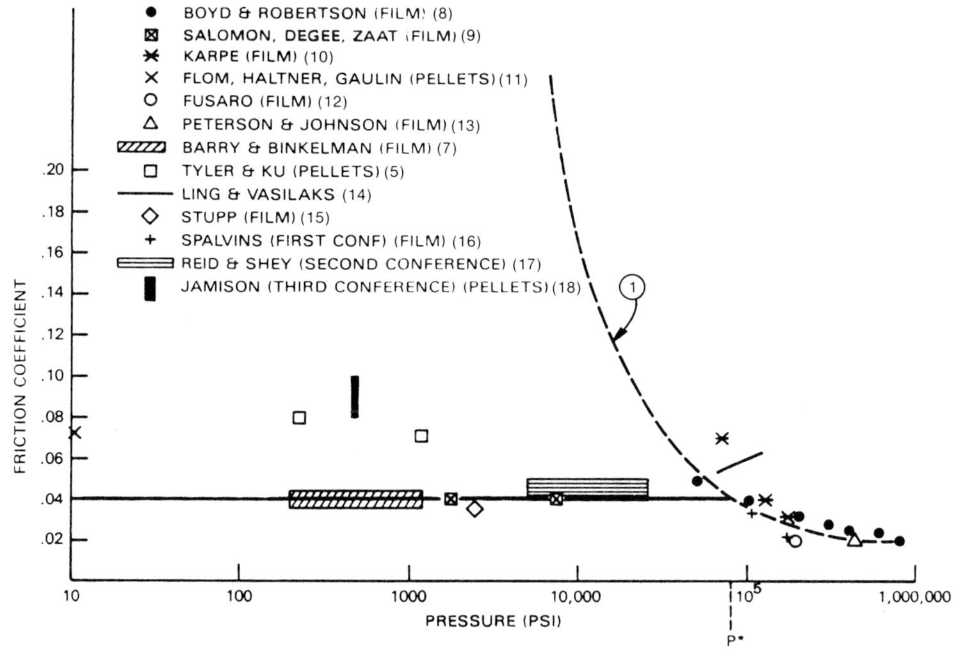

Fig. 1. Effective Pressure for MoS_2 Films.

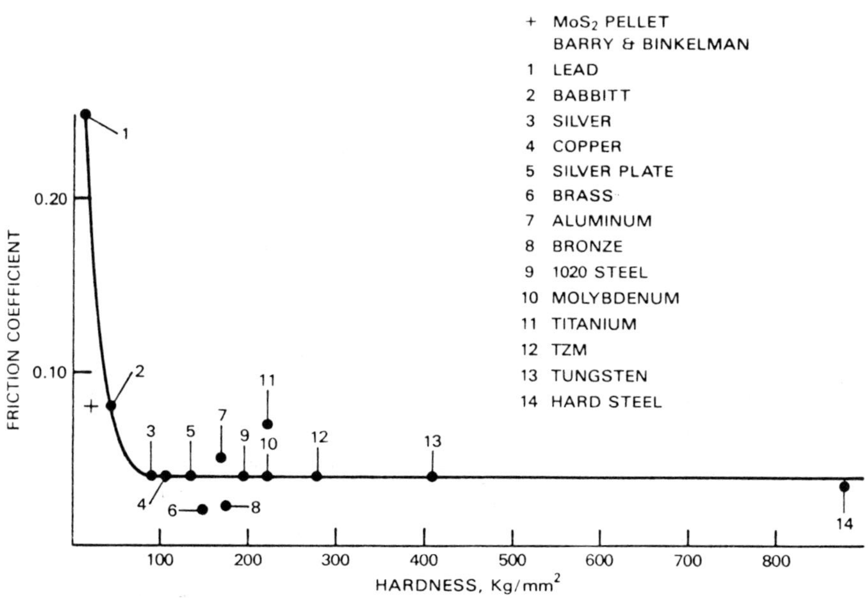

Fig. 2. Effective Substrate Hardness on the Friction Coefficient of MoS_2 Films.

4.9 ON THE ROLE OF ADHESION IN THE WEAR PROCESS

James J. Wert

Tribology research, like that in many other scientific fields, depends upon progress in several different disciplines. Hence, progress in the understanding of friction and wear behavior is somewhat cyclical in nature. Based on limited information, theories are advanced which often cannot be proven or disproved with the experimental capabilities available at the time. With the arrival of each new generation of analytical tools, it is often possible to test existing theories and provide a new framework for understanding fundamental phenomena. Similarly, advances in one discipline can profoundly affect the progress in another.

For example, Kragelsky proposed a fatigue model of wear in 1957 [1]. In his model, repeated stressing of the surface resulted in formation of cracks running parallel to the contacting surfaces. Approximately 15 years later, Suh provided a theoretical foundation based on a dislocation theory [2] and was able to document crack initiation and propagation using scanning electron microscopy.

Kerridge and Lancaster [3] recognized in 1956 the importance of transfer layer formation and stability on the production of loose wear debris. However, understanding the nature of the transfer layer and its role in the wear process has required the availability of modern electron/optical instrumentation such as scanning and transmission electron microscopes, Auger electron spectrometers and x-ray diffraction.

While the role of the transfer layer in determining wear behavior is only partially understood, a great deal of information has been obtained in the past decade. For example, x-ray line broadening studies of Cu-Al alloys [4] have shown that the limiting dislocation cell size developed during sliding contact is a function of the stacking fault energy (SFE) and is similar to the cell size found in the transfer layer. Chen and Rigney [5] have used pin-on-disk sliding tests in vacuum to study transfer layer formation in Cu, Cu-Ni, Ni, and Mo sliding against iron. Detailed studies revealed that of the structure and composition of the wear debris was similar to that observed in a mechanically alloyed powder.

THE ROLE OF ADHESION

Studies have shown that there may be a relationship between transfer layer formation and wear behavior in most adhesive wear processes. However, it is possible that this is a second order effect. That is, while transfer layer formation and stability may be important, the fundamental question is what determines its formation and stability. After all, if transfer layer formation can be prevented, very low wear rates should result.

Recent studies suggest that transfer layer formation is related to the cohesive energies of the contacting solids and the adhesive energy of the interface [6]. The role of adhesion in tribology was first discussed by Desaguliers in 1734. After two centuries of neglect, it was resurrected by Bowden and Tabor in 1939 [7]. With the advances in understanding of surface chemistry and physics, renewed interest in adhesion is warranted and promises to be the most productive area of tribological research for the next decade.

While it is believed that adhesion may be the important parameter in most wear situations, the evidence is often indirect. For example, since the SFE of an alloy is related to ability to cross-slip, the SFE should be related to the delamination theory proposed by Nam Suh [2] and also the fatigue model proposed by Kragelsky [1]. However, when a series of Cu-Al alloys were tested in sliding contact with a sapphire (Al_2O_3) rider, the wear rate increased with decreasing SFE in contradiction of the theory [8]. Buckley, et al. [9] have shown that Al in Cu alloys segregates to the surface under equilibrium conditions and strongly increases adhesion. Therefore, the wear rate increased with increasing aluminum content and hardness and decreasing SFE. When these alloys were tested under abrasive conditions (i.e. surface insensitive), the wear behavior agreed with theoretical predictions.

Schell, Heilman and Rigney [10] conducted similar experiments with Cu-Ni alloys in contact with hardened steel and obtained similar results. They concluded that the effect of SFE was overshadowed by the effect of environment and variations in transfer layer formation. These data suggest that the wear behavior of both the Cu-Al and the Cu-Ni systems is determined by adhesion of the alloys to the riders. Such adhesion determines the formation and stability of the transfer layer.

Further evidence of the importance of adhesion in wear behavior is available in the data obtained by Caldwell [11] for a Al-4.5% Cu age hardening alloy. When 4.5% Cu is added to Al, the wear rate decreases by a factor of 4-5. However, the wear rate was found to be relatively insensitive to both the precipitate morphology (i.e. G.P. zones, θ'', θ', or θ) or bulk hardness. This suggests that the prime role of Cu in modifying the wear behavior of Al is to decrease the adhesive strength of the alloy to the rider.

From the fundamental viewpoint, adhesive wear must be related to the cohesive strengths of the two counterfaces and the adhesive strength of the interface. Based on the cohesive energies of the metals involved, Buckley [9] was able to predict that the direction of material transfer would be from the cohesively weaker to the cohesively stronger material. Through a series of elegant experiments, using Auger analysis, LEED and a sophisticated loading system, he was able to verify theoretical predictions and also measure the adhesion force of several metals in contact with themselves and also with Cu, Au Fe and several nonmetals such as Al_2O_3 and SiC. He found that in both pure metal couples and metal/nonmetal couples, that the adhesive force was dependent on the chemical reactivity and the degree of d-bonding.

In close packed metals such as fcc copper, the atoms of the bulk have 12 nearest neighbors. However, at a free surface the number of nearest neighbors, i.e. the coordination number, is dependent upon the atomic plane being considered. The coordination number for the (111) surface plane is reduced from 12 to 9 and gives rise to a surface energy of ~2500 ergs/cm^2. As the density of packing decreases, the surface energy increases due to fewer near neighbors being available to attract the valence electrons. For the (100) plane of Cu, the coordination number is reduced from 8 to 7 and results in a surface energy of ~2900 ergs/cm^2. Because of the higher surface energy, the (100) planes are more chemically reactive and yield higher adhesive forces when in contact with other materials. Buckley [9] has measured the adhesive force of Cu single crystals in contact with a (100) surface of Au. He found the adhesive forces developed by the (111), (100) and (110) planes to be 80, 185 and 390 mg, respectively. That is, the adhesive force varied inversely with the coordination number.

Since the density of atomic packing and coordination number are dependent on crystal structure it is expected that adhesion should also be related to crystal structure. Buckley [9] has found the adhesion coefficient of the (0001) basal plane of cobalt in contact with another basal plane to be only 0.05. Such a low value implies that fracture would occur along the original interface and result in a low coefficient of friction and wear rate. Similar results were obtained with fcc Cu in contact with bcc tungsten.

SUMMARY

Based on the information available, it is believed that the interaction between the mechanical properties and adhesive characters of the contacting surfaces determines the wear behavior of many tribosystems. Therefore, any fundamental theory of wear behavior must include directly or indirectly, adhesive and cohesive energies, mechanical properties and structure of the contacting surfaces.

REFERENCES

1. *Friction and Wear*, I. V. Kragelsky, translated by N. Standen, Pergamon Press, Inc., Elmsford, NY (1982).
2. "The Delamination Theory of Wear," N. P. Suh, *WEAR*, *25*, 111-123 (1973).
3. "The Stages in a Process of Severe Metallic Wear," M. Kerridge and J. K. Lancaster, *Proceedings of the Royal Society (London)*, *A236*, 250-264 (1956).
4. "An X-ray Diffraction Study of the Effect of Stacking Fault Energy on the Behavior of Cu-Al Alloys," J. J. Wert, S. A. Singerman, S. G. Caldwell and D. K. Chaudhuri, *WEAR*, *92*, 213-229 (1983).
5. "Transfer During Unlubricated Sliding Wear of Selected Metals Systems," L. H. Chen and D. A. Rigney, *WEAR*, *105*, 47-61 (1985).
6. "The Influence of Stacking Fault Energy and Adhesion on the Wear of Cu and Al-Bronze," J. J. Wert and W. M. Cook, to be published in *Wear of Materials 1987*, American Society of Mechanical Engineers, Houston, TX (April 1987).

7. "The Area of Contact Between Stationary and Between Moving Stresses," F. P. Bowden and D. Tabor, *Proceedings of the Royal Society (London)*, *A169*, 391 (1939).
8. "The Effects of Stacking Fault Energy on the Sliding Wear Behavior of Cu-Al Alloys," S. G. Caldwell, J. J. Wert and R. W. Carpenter, *Wear of Materials 1981*, American Society of Mechanical Engineers, New York, 65-68 (1981).
9. *Surface Effects in Adhesion, Friction, Wear and Lubrication*, D. H. Buckley, Oxford University Press, London (1981).
10. "Friction and Wear of Cu-Ni Alloys," J. Schell, P. Heilman and D. A. Rigney, *Wear of Materials 1981*, American Society of Mechanical Engineers, New York, 53-62 (1981).
11. "Dry Sliding Wear of Selected Two Phase Al-Cu Alloys," S. G. Caldwell and J. J. Wert, *Wear of Materials 1987*, American Society of Mechanical Engineers, 1987 International Wear Conference, Houston, TX (April 1987).

4.10 POSSIBLE RELATIONSHIP BETWEEN PARTIAL-ELASTOHYDRODYNAMIC LUBRICATION AND WEAR MODELLING

Herbert S. Cheng

INTRODUCTION

In most lubricated contacts, the lubrication mode is partially elastohydrodynamic, that is, a significant portion of the load is carried by the asperity contacts. In this regime, wear phenomenon is influenced by how the asperities are protected by the lubricant or surface film during sliding.

A realistic modelling of the wear of a single asperity contact and eventually the wear rate of the entire lubricated contact should include the possible influences of the relevant lubrication variables on the wear rate. It is with this objective the following discussions are written.

ASPERITY LUBRICATION AND WEAR

When a hard lubricated asperity slides across an opposing surface, as shown in Fig. 1, materials of the soft surface can be plastically extruded out the surface and eventually detached from the surface to form wear debris. This type of wear is commonly recognized as two-body abrasion or plowing wear, which is dependent upon the asperity slope and interfacial friction. For the case of very small asperity slope and very large interfacial friction, the wear debris is largely formed by the interfacial shear stress; this process is more closely related to adhesion than abrasion. If both the asperity slope and the frictional stress are small, then the deformation would be mainly elastic. For this case, material is generally removed by pitting resulted from surface and subsurface crack propagation and fracture. Aside from the above described mechanical wear modes, materials can also be removed through chemical reaction with the environment or the lubricant, such as oxidation wear or corrosive wear. In the presence of lubricant, the removal of metal by metal/lubricant reaction or by diffusion from one metal to another is classified as chemical wear.

In one way or another, the wear modelling is influenced by the lubrication conditions at each asperity. Before proceeding to discuss the role of the lubrication variables in wear, a brief description is given to partial-elastohydrodynamic (EHL) lubrication and asperity lubrication.

PARTIAL-EHL AND WEAR MODELLING

In the film-EHL, the important parameters are lubricant film thickness, pressure, shear stress, film temperature, and surface temperatures. These variables are still carried over to partial-EHL except that in partial-EHL each variable now contains local fluctuations caused by the presence of asperities and valleys. These local variations in partial-EHL are treated as random variables within an elemental bearing area which contains a sufficiently large number of asperities. From the point of view of wear modelling, it is the local variables at the asperities and their distributions which are important.

Fig. 2 shows qualitative sketches of possible distributions of the asperity film thickness, contact pressure, shear stress and contact temperature. Obviously, wear modelling is more strongly related to the thinner asperity film thickness, higher asperity pressure, shear stress, and contact temperature. One might start the modelling from the wear analysis of a single lubricated asperity, and then put it together to yield a wear rate prediction in the entire contact area.

Due to the fact that wear mechanisms are multi-faceted, it is very unlikely that a unified wear theory might be found to calculate the wear rate as a function of the materials and lubricant properties for all wear modes. For modelling of mechanical wear, the most logical starting point appears to be the plasticity analysis of the soft substrate slid by a hard conical or spherical asperity with the effects of surface film lubrication. This work involves a three-dimensional plasticity analysis. It is clearly difficult but does not appear to be unattainable. By varying the conical tip geometry, this analysis should shed considerable light on all three major modes of mechanical wear mentioned earlier. The role of asperity pressure and asperity shear stress in this plasticity analysis is clearly evident. The asperity temperature has no direct effect on plastic deformation, but it can have an indirect influence because of its effect on hardness and asperity shear stress.

In modeling chemical wear, the asperity temperature plays a primary role because of its control on the chemical reaction rate. However, the asperity pressure, shear stress, and film thickness are equally important in chemical wear because of their strong influence on asperity temperature.

CONCLUDING REMARKS

From the present understanding of the wear mechanisms, modelling of lubricated wear based on sliding lubricated asperities would require almost all pertinent variables in partial-elastohydrodynamic lubrication whether the wear mode is mechanical or chemical. The foundation of wear modelling appears to rest on a three-dimensional plasticity analysis of a sliding lubricated asperity, which to date has not been achieved. This analysis requires the input from micro-EHL. In modelling of wear rate in a lubricated contact containing many asperity encounters, one would need to know the average film thickness, number of asperity contacts per unit area, and statistical distributions of asperity contact pressure, shear stress, and temperature. All of these must come from partial-EHL analysis.

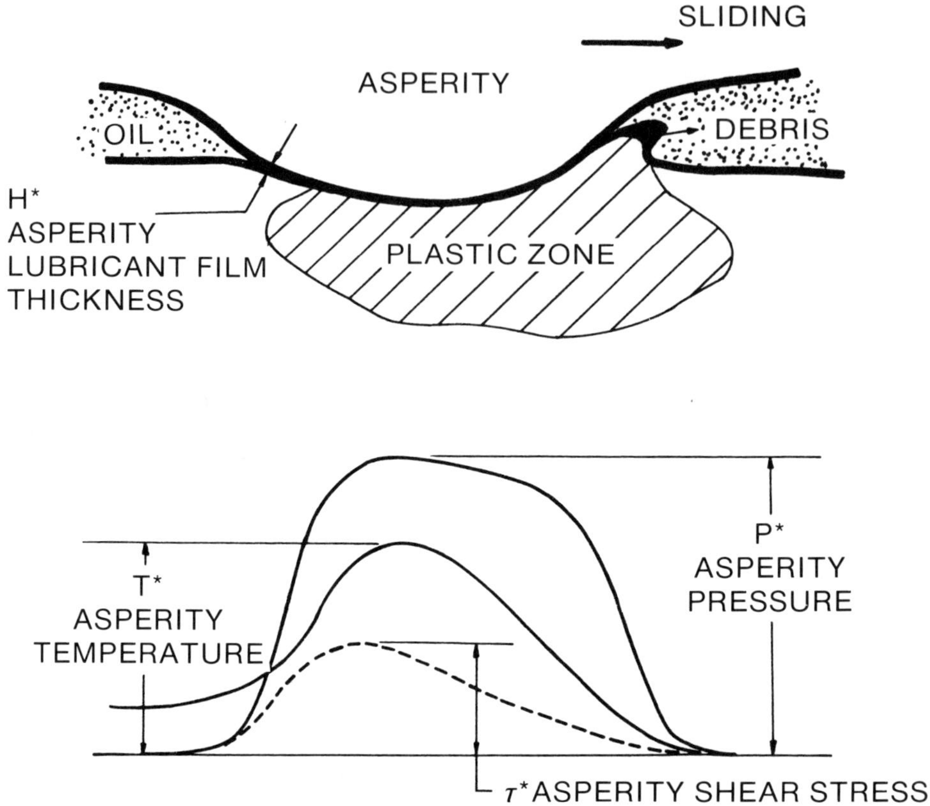

Fig. 1. The Asperity Film Thickness, Pressure, Shear Stress, and Temperature at the Sliding Interface Between a Hard Asperity and a Soft Substrate.

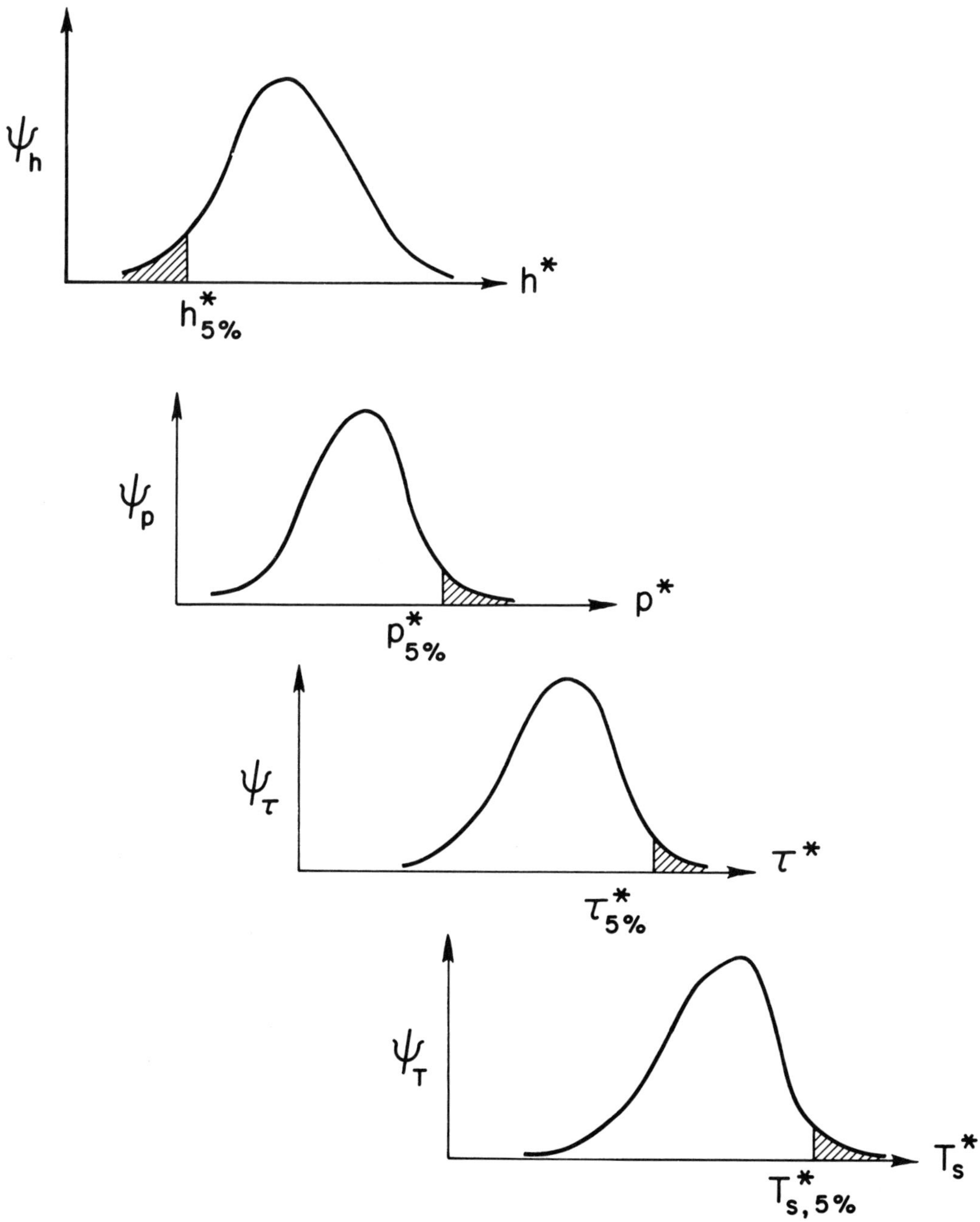

Fig. 2. Typical Distributions of Asperity Film Thickness, Pressure, Shear Stress, and Temperature.

4.11 SLIDING SYSTEMS WITH NO WEAR

Norman S. Eiss, Jr.

INTRODUCTION

There are certain systems in which the primary function of sliding contacts is not the transmission of motion or work. For example, the purpose of a sliding electrical contact is to provide a low resistance electrical path between two bodies in relative motion. The contact between the magnetic head and a floppy disk or a video cassette recorder tape is necessary to maximize the magnetic field stored in the magnetic domains in the coating on the disk or tape. In these systems, the normal load at the contacts is minimized within the other constraints of each system. One purpose for minimizing the normal load is to minimize wear.

Wear debris in sliding electrical contacts can have two detrimental consequences: increasing contact resistance and shorting out adjacent circuits. Wear of the heads in magnetic storage systems eventually destroys the ability of the head to write and read data and wear of the media, which is almost always catastrophic, causes a loss of stored information. Consequently, in both systems the desired situation would be no wear.

Wear models can be used to identify parameters and conditions which could be varied to minimize wear. Rabinowicz has proposed that systems with wear coefficients of 10^{-6} to 10^{-8} are experiencing burnishing wear [1]. The mechanism of burnishing wear is assumed to be a process of single-molecule removal from the peaks of the asperities. Challen and Oxley [2] use a slip line model of a plastically deformed wave of material pushed ahead by an asperity to predict the wear coefficient K. Their model is based on the total plastic shear strain required to produce a wear particle where the total strain is the accumulation of n cycles of incremental shear strain per pass. Their model shows that low values of K result for asperities with small slopes and low values of the ratio of the shear strength of the interface to the shear flow stress of the plastically deformed material. An empirical approach for predicting low wear was developed by Bayer et al. [3] in which wear on the order of the original surface finish of the material was called "zero wear." While these methods were developed to predict low rates of wear, they are of little use in predicting the sliding distance which can be experienced with no measureable wear prior to the initiation of a catastrophic high rate of wear.

In this paper, some experimental studies of the no wear-catastrophic wear phenomenon will be summarized and some suggestions will be made for both experimental and modeling of this phenomenon.

EXPERIMENTAL OBSERVATIONS

Systems in which no wear-catastrophic wear occurs have involved polymeric-based materials in an experiment in which a polished steel ball is slid on the polymeric surface. The load is selected so that when sliding commences there is no wear occurring on the polymeric surface. In all cases the track of the ball on the polymer is visible to the naked eye which indicates that there has been a modification of the surface topography of the polymer, i.e., a slight flattening of the contacting asperities. After a certain duration of sliding, a wear groove forms. The initiation of the wear groove is accompanied by a sharp rise in the coefficient of friction, the generation of wear particles, and the formation of the groove which has a circular segment as its cross-sectional area. The wear rate after the initiation of the wear groove is linear with sliding distance.

Polymeric systems in thich this phenomenon has been observed include polycarbonate [4], polyvinyl chloride [4], several polyimides [5,6], epoxies and rubber modified epoxies [7,8], and magnetic coatings (consisting of magnetic particles, organic binders, and alumina particles) on rigid disks [9].

Some attempts have been made to study the change in the sliding track prior to the initiation of the wear groove. In these studies, tests were performed to measure the number of cycles of disk rotation required to initiate wear. In subsequent tests, the experiments were stopped before initiation and the sliding track was examined using optical and scanning electron microscopy. In epoxies, for example, the first change observed was the appearance of small cracks perpendicular to the sliding direction and located in the center of the sliding track. These cracks are similar to those observed when a blunt body is slid on glass [10]. These cracks are caused by a tensile stress field which trails the slider. Because these cracks are not formed immediately when sliding commences, they may result from the propagation of subsurface cracks to the surface or from nucleated subsurface voids.

The surface cracks increase in length as a result of subsequent stress cycles. More cracks appear and eventually a loose wear particle is formed. Usually the loose wear particle is trapped between the slider and the surface and dragged along the surface plowing a groove and loosening more wear particles from the cracked surface. After more cycles of sliding, additional plowed grooves are formed by particles being dragged by the slider. At some time subsequent to these initial plowed grooves, the wear spreads over the entire central region of the sliding track and a groove is formed in the surface.

For two-phase materials, such as rubber-modified epoxy materials in which the rubber modifier segregates into domains during curing, the wear groove initiation is modified from that observed for the pure epoxy. The elastomeric material in the domains is roughened. At the location where the friction force causes a tensile stress between the elastomer and the matrix, separation occurs and on the opposite side of the domain the elastomer may be extruded onto the surface of the epoxy.

The presence of these elastomeric domains affects the generation of wear particles from the epoxy matrix in three ways. First, the presence of the domains reduces the elastic modulus of the material. Hence, for a given load the area of contact will be larger and the stresses lower. These lower stresses should prolong the duration of the initiation of the wear track. Second, if the spacing between the domains is small, the cracks appear in the epoxy matrix much sooner. This early appearance of cracks is probably caused by the higher stresses created in these epoxy column-like regions between the elastomeric domains. Third, occasionally a ring of small cracks appears around an elastomeric domain. These cracks could represent a mechanism of energy dissipation which may then delay the initiation and propagation of cracks which lead to wear particles and wear groove formation.

DISCUSSION

If one wishes to prolong the duration before the wear groove forms, then steps must be taken to prevent voids from forming and cracks from propagating. If the material has a hard inclusion, then the deformation of the matrix around the inclusion must be limited so that the strength of the bond between the materials is not exceeded. In modeling void formation around hard inclusions in metals, Suh [11] has postulated that when the stored elastic energy around the inclusions is less than the surface energy of the potential void there is a particle size below which voids will not form even if the stress exceeds the bond strength. Hence, one step to take would be to minimize the size of inclusions. In a heavily filled material like a magnetic coating, this step is not possible. Therefore, one must try to increase the strength of the bond between the particles and the matrix and decrease the stress at the interface by controlling the shape of the particles and the magnitude of the normal stress.

The cracks which formed in the epoxy matrix appear to be propagated by the tensile stress which trails the slider. One function that the elastomeric domains may have in prolonging the initiation of the wear groove is the reduction in the traction between the slider and the rubber modified epoxy. The modifier is a siloxane polymer which has a relatively low surface energy and also has a higher concentration at the free surface when curing. In materials which have cracks and are subjected to nominal elastic stresses, there is a threshold stress intensity which must be exceeded before the stress will cause the cracks to grow [12]. If existing cracks are to be prevented from growing, the stresses must be limited. This can be accomplished by reduction in normal load, reduced shear strength at the sliding interface, and by decreasing the elastic modulus of the material.

In all of the studies described above the observations have been limited to optical and scanning electron microscopy of the surface before and after the wear groove formed. More information is needed on subsurface changes during the initiation period. Techniques for observing the subsurface by sectioning and use of higher magnifications available in the scanning transmission microscopic must be developed, particularly for the multiphase materials such as magnetic coatings.

It must be noted that the phenomenon of no wear-catastrophic wear occurs in systems in which the polymer is brittle. For more ductile polymers wear under low stress appears to involve transfer and the wear rate depends on the adherance of the transferred material to the slider. The catastrophic development of a wear groove is not observed for these polymers.

CONCLUSIONS

In certain systems where sliding contact occurs the desired tribological phenomenon is no wear. In systems in which a brittle polymer is one of the sliding components, a duration of no wear can be followed by a sudden initiation of a wear groove and its subsequent enlargement. The mechanism of the initiation of this wear groove is not well understood and observations of subsurface changes prior to the wear groove initiation are required particularly in magnetic coatings which have a large volume fraction of hard particles in the polymer matrix. High magnification of the subsurface region using scanning transmission microscopes would be desirable. Models of the initiation of the wear track are needed. These models must include the stresses at particle matrix interfaces and crack propagation in the matrix in the vicinity of the hard particles.

REFERENCES

1. "The Least Wear," E. Rabinowicz, *WEAR*, *100*, 533-541 (1984).
2. "Prediction of Archard's Wear Coefficient for Metallic Sliding Friction Assuming a Low Cycle Fatigue Wear Mechanism," J. M. Challen and P. L. B. Oxley, *WEAR*, *111*, 275-288 (1986).
3. "Designing for Zero Wear," R. G. Bayer, A. T. Shalkey and A. R. Wayson, *Machine Design*, *41*, 142-151 (1969).
4. "Fatigue Wear of Polymers," N. S. Eiss, Jr. and J. R. Potter, III, in *Polymer Wear and Its Control*, L.-H. Lee, editor, American Chemical Society, Washington, DC, 59-66 (1985).
5. "Effect of Chemical Structure on the Friction and Wear of Polyimide Thin Fibers," J. W. Jones and N. S. Eiss, Jr., in *Polymer Wear and Its Control*, L.-H. Lee, editor, American Chemical Society, Washington, DC, 135-146 (1985).
6. "Friction and Wear of Polyimide Thin Films," M. Chitsaz-Zadeh and N. S. Eiss, Jr., *WEAR*, *110*, 359-368 (1986).
7. "Siloxane - Modified Epoxies," E. M. Yorkgitis, N. S. Eiss, Jr., C. Tran, G. L. Wilkes and J. E. McGrath, in *Advances in Polymer Science*, *72*, *Epoxy Resins and Composites I*, K. Dusek, editor, Springer-Verlag, Berlin, 79-110 (1985).
8. "Siloxane Modifiers for Epoxy Resins," E. M. Yorkgitis, C. Tran, N. S. Eiss, Jr., T. Y. Hu, I. Yilgor, G. L. Wilkes and J. E. McGrath, in *Advances in Chemistry, Series 208*, *Rubber-Modified Thermoset Resins*, C. K. Riew and J. K. Gillham, editors, American Chemical Society, Washington, DC, 137-162 (1984).
9. "Spherical Pin Sliding Test on Coated Magnetic Recording Disks," Y. Kawakubo, H. Ishihara, Z. Tsutsumi and J. Shimizu, in *Tribology and Mechanics of Magnetic Storage Systems*, *III*, B. Bhushan and N. S. Eiss, Jr., editors, American Society of Lubrication Engineers, Park Ridge, IL, 118-124 (1986).
10. "Review - Indentation Fracture: Principles and Application," B. Lawn and R. Wilshaw, *Journal of Materials Science*, *10*, 1049-1081 (1975).
11. *Tribophysics*, N. Suh, Prentice Hall, Inc., Englewood Cliffs, NJ, 141-147 (1986).
12. *Ibid.*, 153-157.

4.12 COMPARISON OF WEAR CHIP MORPHOLOGY WITH DIFFERENT MODELS OF "ADHESIVE" WEAR

Yu Jun Chang and Doris Kuhlmann-Wilsdorf

At least six basic models of wear chip formation in unlubricated sliding between ductile materials exist, as indicated in Fig. 1. These are (a) breaking or shearing-off pieces of the softer material where asperities of the two sides collide (Fig. 1A); (b) shearing-off flakes of the softer material where local adhesion has occurred at contact spots (Fig. 1B); (c) plucking out fragments of the softer material where strong adhesion, such as through cold-welding, has created a firm bond between the two sides at contact spots (Fig. 1C); (d) shearing-off of flakes wherein the relative rest between the two sides at the contact spots is not due to adhesive forces but to mechanical micro-roughness interlocking (Fig. 1D); (e) flaking of a debris layer at contact spots (Fig. 1E); (f) "Delamination" of flakes through sub-surface cracks parallel to the interface which spread through repeated passes of asperities (Fig. 1F). "Wedge" or "Prow" formation is believed to occur through repeated action according to model (b).

Wear particles formed through mechanisms (a) and (d) will not be bound to the harder side by "adhesive" forces, meaning forces with a significant component normal to the interface. These are therefore liable to fall free readily if that is geometrically possible. Also wear particles formed as in Figs. 1E and F may fall free in the same manner, provided that the relative rest between the two sides at the interface was enforced through micro-interlocking rather than through adhesion. However, if adhesion plays an important role in models (e) and (f), and certainly in models (b) and (c), the wear fragments will initially adhere to the harder side and will become free only after a force has been applied normal to the interface which is at least equal to the adhesive force.

Clearly, for the condition directly after the detachment of the particles, the models indicated in Fig. 1 predict characteristically different features for all surfaces, including the two sides of the wear particles and of the sliding materials, and also predict characteristic wear fragment shapes. These characteristic distinguishing features, which will be summarized presently, could therefore be used to ascertain which of the modelled mechanisms operates in cases of interest, except that uncontrollable damage tends to obliterate them almost immediately. This occurs because the free gap between the two sides outside of the contact spots is normally narrower than the thickness of the wear particles. As a result these become deformed, fragmented, oxidized, coagulated together and/or reattached to either of the two sides. Indeed, the so commonly observed debris layer as in Fig. 1E is the result of those processes.

Such subsequent damage is minimized when using samples which are so small as to accommodate only one or a very few contact spots on the average, since in that case the wear chips make their way out of the gap before suffering too severe damage and, conversely, thereby damage the two sides or may become reattached. Evidently, thin wires or fibers moving on a substantially flat substrate fulfill the above requirement. Further, in order to generate a sufficiently large number of wear particles (or "chips") without too long waiting times, it is obviously necessary to use many wires or fibers, i.e. to use somewhat loosely formed fiber bundles. This was done for the case of copper fibers of different diameters sliding against a smooth copper rotor under a variety of different sliding conditions [1].

The expected chip morphologies for the models of Fig. 1 are these: More or less equiaxed fragments in cases (a) and (c). Mildly elongated particles whose somewhat irregular thickness is typically less than their length and width in cases (b) and (d). Platelets with relatively uniform thickness in cases (e) and (f), these consisting of "debris" intermixed with oxides and perhaps still other surface film material in case (e) but mostly of the pure softer material in case (f).

As to surface morphology, that side of the chip which initially faced the harder side will be puckered, i.e. will show the typical structure of ductile fracture in the case of chips formed by mechanisms (b) and (c), and fragments of the wear particles will remain attached to the harder side. Those features are also expected if adhesion instead of micro-roughness interlocking operated in cases (e) and (f). Conversely, models (a) and (d), as well as (e) and (f) in the absence of adhesion, predict wear particles which, on the initial interface side, still retain traces of micro-roughness. That roughness is formed through the unavoidable inhomogeneities of plastic deformation (see Ref. 2) and has an expected profile depth on the order of 1000Å.

The wear fragment face originally turned away from the interface has been simply sheared off in cases (a), (b) and (d) and should appear accordingly, while in case (c) it should be puckered through ductile fracture. Again, in case (e) the surface initially facing away from the interface basically decohered and should therefore mirror the surface morphology of the substrate where the debris layer flaked off. Lastly, in case (f), it should be one side of a type II crack that grew in stages and therefore might exhibit the corresponding terracing.

In all cases, the surfaces exposed on the softer side where the wear fragment was detached should mirror the morphology of the corresponding chip surfaces just discussed. The harder side, on the other hand, should mirror the morphology of the contact spots. At the same time, there are adhering wear chips and small fragments if strong adhesion takes place, as in cases (b) and (c) and perhaps in cases (e) and (f). Finally, in all cases only partly detached wear chips should occur and be seen on the softer side. These would have the appearance of "tongues" overlying the softer side in cases (b) and (d) (see the middle sketches to Figs. 1B and D), have the appearance as of platelets in the process of peeling off the softer side in cases (e) and (f), of irregular blobs still attached to the softer side in case (a) and also case (c) if particle formation is still in the first phase, and finally of roughly equiaxed particles dangling from the harder side in the latter stage of case (c).

It shall not be doubted that all of the above mechanisms can operate, depending on specific conditions. In the already cited study of bundles of fine copper wires sliding against a copper substrate [1], the result unequivocally supported case (d) with all of the predicted features indeed observed. Here the fibers represented the softer side. On these, the discussed "tongues" of partly formed wear chips as in the middle sketches of Figs. 1B and D were found, and on the opposing side nearly round contact spots. The shape and size of the contact spots on the harder side closely matched the flattened areas on the "tongues," i.e. those parts of the "tongues" which had been interlocked with the contact spots on the harder side. In turn, these areas matched the shape and size of the wear chips. Also, careful examination revealed instances in which the micro-roughness could still be discerned, on chips as well as on the contact spots.

Additional confirmation of the operation of model (d), and this for the particular case of a fixed number of contact spots per fiber end, was obtained from a quantitative analysis of the shape and size of the chips as a function of applied load, P: As expected from simple theory for constant number of contact spots [1-4], the average load-bearing area per contact spot, and thus the average area of the wear fragments, was found to be proportional to P, as seen in Fig. 2, while the length to width to thickness ratio remained constant (Fig. 3).

ACKNOWLEDGEMENT

One of the authors (D. K.-W.) gratefully acknowledges the financial support of this research through the Materials Division (Tribology) of the Office of Naval Research, Arlington, VA; the other author (Y. J. C.) that of the Shanghai Iron and Steel Research Institute, Shanghai.

REFERENCES

1. "A Case of Wear Particle Formation Through Shearing-Off at Contact Spots Interlocked Through Micro-Roughness in 'Adhesive' Wear," Y. J. Chang and D. Kuhlmann-Wilsdorf, *Wear of Materials - 1987*, K. C. Ludema, editor, American Society of Mechanical Engineers, New York (1987), in press.
2. "Dislocation Concepts in Friction and Wear," D. Kuhlmann-Wilsdorf, *Fundamentals of Friction and Wear of Materials*, D. A. Rigney, editor, American Society of Metals, Metals Park, OH , 119 (1981).
3. *Electric Contacts*, R. Holm, Springer-Verlag, Berlin/Heidelberg/New York, 4th edition (1967).
4. "Contact and Rubbing of Flat Surfaces," J. F. Archard, *Journal of Applied Physics*, *24*, 981 (1953).

LEGENDS

Fig. 1: Schematic diagrams for six different models of wear chip formation. The harder material is at top, the softer at bottom. A horizontal dashed line indicates adhesive bonding. The models are, in order:

(A) Interference of asperities.

(B) Shearing-off at contact spot at which relative rest is enforced through adhesion. The wear particle detached from the softer side clings to the harder side until it is detached by some unspecified force with a normal component away from the harder side of a strength at least equal to the adhesive force. That will pucker the top side as in (C) below.

(C) As (B) but detachment of the particle through a plucking-out action rather than shearing off. This is the model apparently used by Archard [4]. The wear particle is roughly hemispherical rather than a flake as in (B).

(D) As (B) but with micro-roughness interlocking at the contact spot acting instead of adhesion. As a result the wear particle falls free easily.

(E) As either (B) or (D) but acting on a debris layer. The wear particle will be of a much more uniform thickness than in (B) and (D) and will, of course, be composed of debris rather than the pure material of the softer side.

(F) Delamination, i.e. detachment of a particle through the nucleation and growth of a subsurface crack more or less parallel to the interface, due to the action of repeated passes of a contact spot.

Fig. 2: Average area of wear chips obtained from dry-sliding of a bundle of 178μm thick copper wires run end-on against a copper rotor, as a function of applied load. The wear fragments were collected on microscope glass slides directly next to the exit side of the wire bundle/rotor interface. They were analyzed by means of a Lemont Scientific Image Analyzer. Since one expects a fixed number of contact spots per fiber end, simple theory predicts proportionality between average area and applied force for model (d), as observed.

Fig. 3 Ratio of average thickness to length $(\overline{t}/\overline{\ell})$ and width over length $(\overline{w}/\overline{\ell})$ values of wear chips obtained from wire bundles of different diameters as in Fig. 2. As expected from model (d), these ratios are (almost) independent of applied force, P. The mild decrease with P is believed to be due to chip breakage which mostly shortens the fragments and becomes more frequent as load and chip size increase.

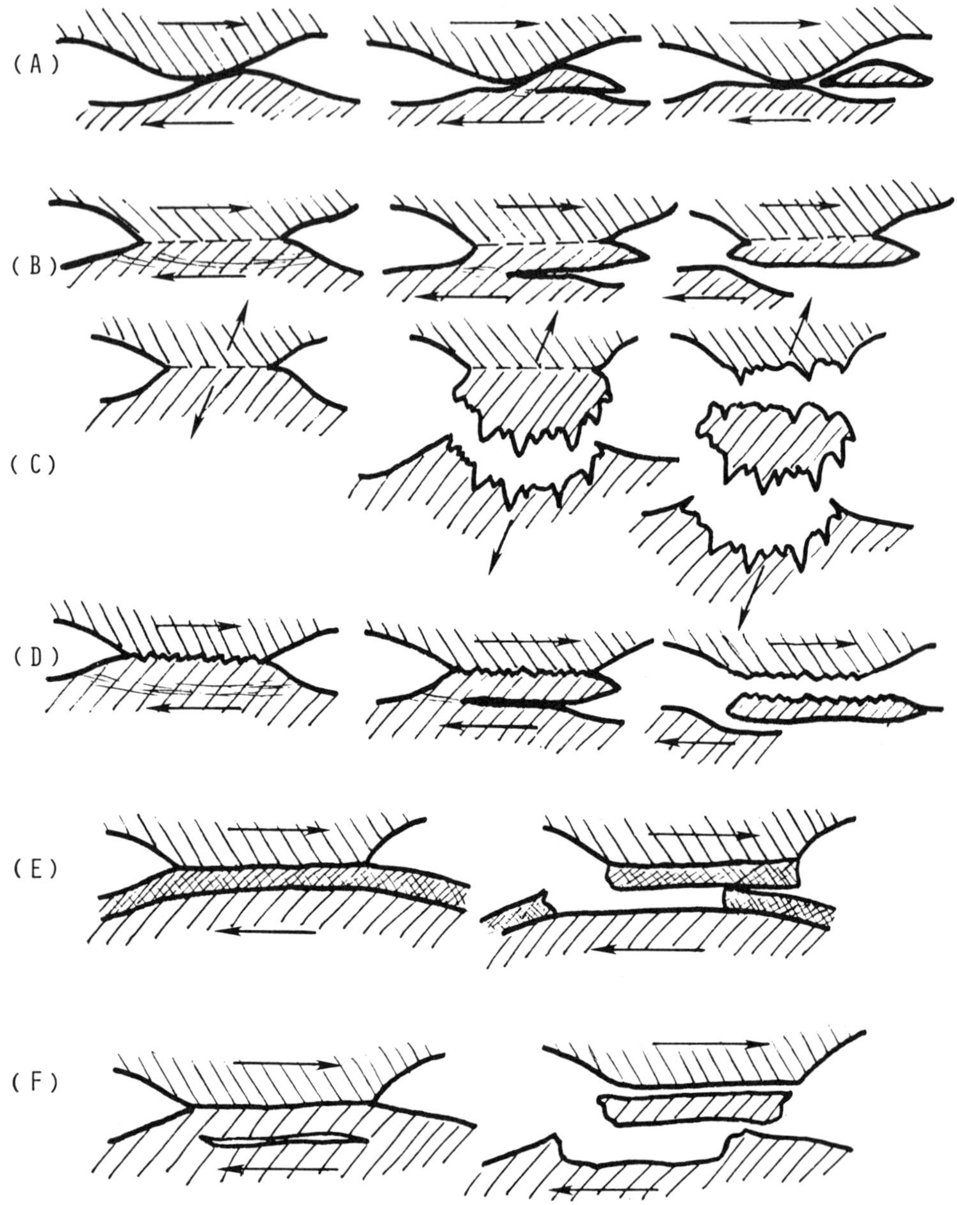

FIG:1

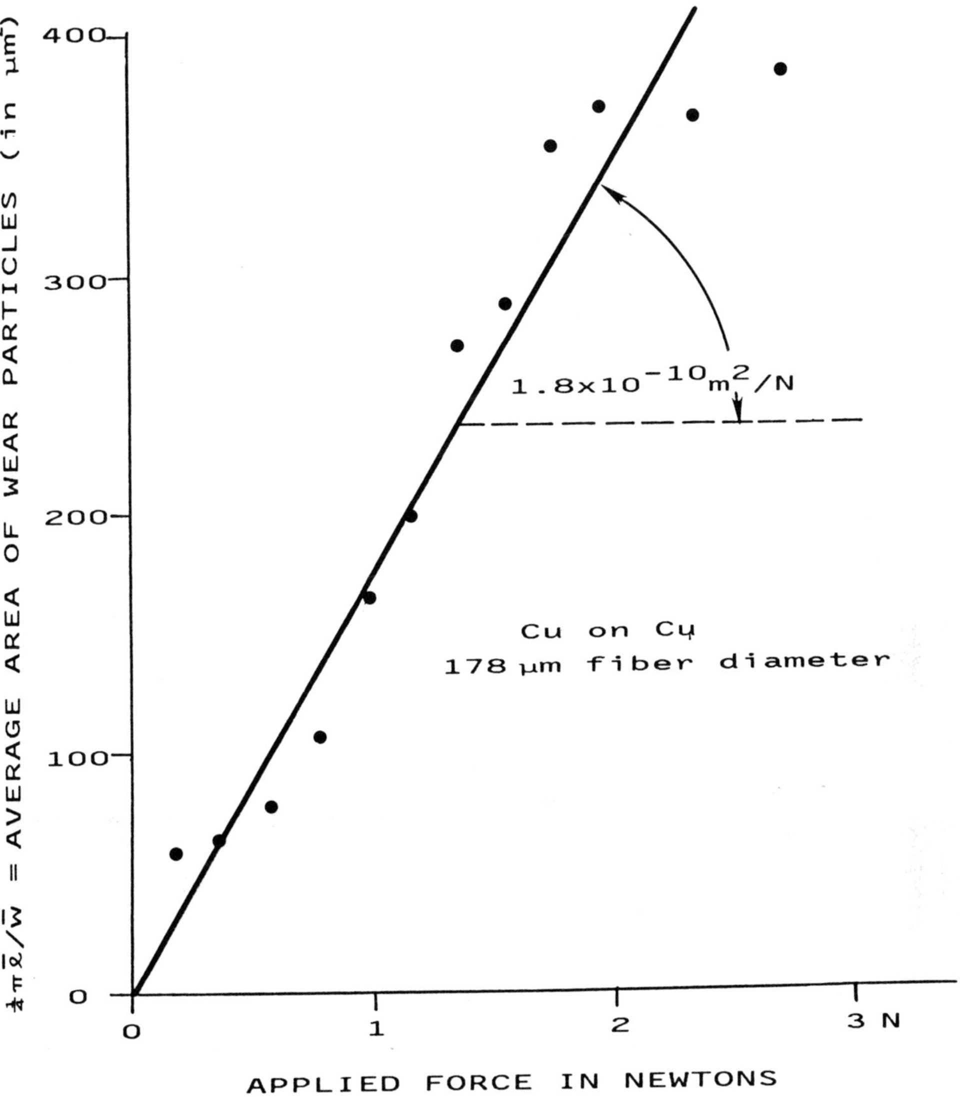

$\frac{1}{4}\pi\bar{l}/\bar{w}$ = AVERAGE AREA OF WEAR PARTICLES (in μm^2)
400
300
200
100
0
0
1
2
3 N
APPLIED FORCE IN NEWTONS
$1.8\times10^{-10} m^2/N$
Cu on Cu
178 µm fiber diameter

FIG:2

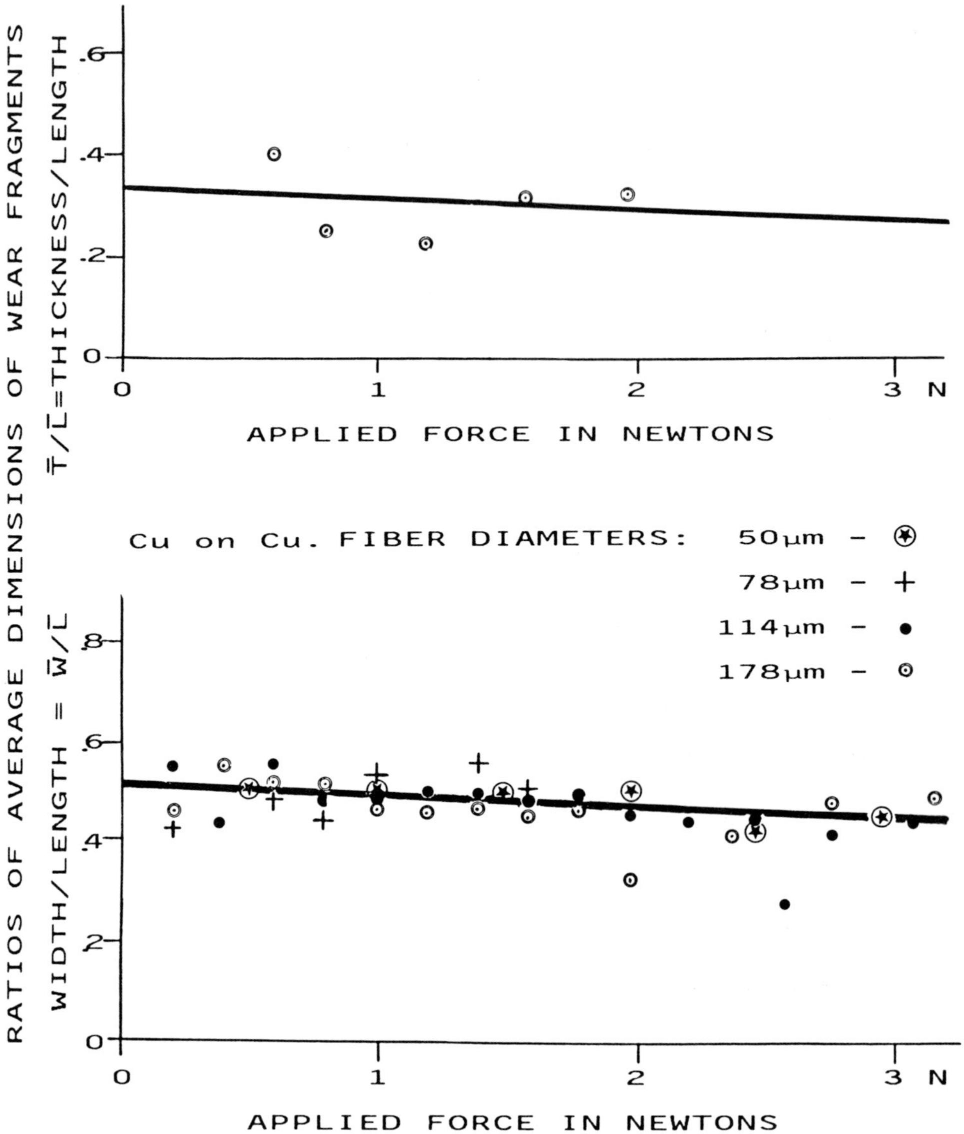
RATIOS OF AVERAGE DIMENSIONS OF WEAR FRAGMENTS
T̄/L̄=THICKNESS/LENGTH
.6
.4
.2
0
0
1
2
3 N
APPLIED FORCE IN NEWTONS
Cu on Cu. FIBER DIAMETERS:
50μm
78μm
114μm
178μm
WIDTH/LENGTH = W̄/L̄
.8
.6
.4
.2
0
0
1
2
3 N
APPLIED FORCE IN NEWTONS

FIG:3

4.13 A PROPOSED THERMOMECHANICAL WEAR THEORY

Bond-Yen Ting and Ward O. Winer

INTRODUCTION

The transmission of load and motion in a mechanical system rely on the relative contact movement between the tribological elements. However, tribological contacts induce surface tractions and wear takes place. Although wear processes in tribo-contacts are still not fully understood, a major cause of wear is the interaction between the asperities of the contact surfaces. Various types of wear mechanisms have been considered [1,2]. Adhesion, abrasion, fatigue and corrosion of the material are some of the important wear mechanisms.

An important factor has long been ignored in the development of wear theories -- the thermal effect from frictional heating, which is always coexistent with tribo-contacts. When relative motion exists between two contacting surfaces, the work imposed on the contact area to overcome the friction is transformed into thermal energy at the interface. As a consequence of transporting this interfacial thermal energy, the contact temperature rises, and a thermal stress field is created. For any dynamic contact between surfaces, in addition to the stress field generated by the surface tractions, the thermal stress field has to be considered. The temperature rise may also change the material properties at the contacting asperities and soften the material. The frictional heating, therefore, can have multiple influences on the contact and can be important to the wear process.

The tribo-contact temperature studies of Blok [3] and later Archard [4] first drew attention to the importance of thermal phenomena in tribo-contacts. Barber [5,6,7,8,9,10] and Burton [11,12,13] and co-workers studied the thermoelastic instability and the surface distortions due to the thermal expansion from frictional heating. Ling [14] showed that the contact temperature influences the seizure of metal contacts, and used both a statistical model and a dynamic thermocouple method to determine the contact temperature [15,16]. Kennedy and Ling [17] employed a finite element method to calculate the stress field and used maximum distortion energy yield criterion to predict the wear of brake pad. Quinn [18,19,20,21] proposed an oxidational wear theory in which the contact temperature is included to find the surface oxide film thickness, and determine wear from the detachment of the oxide film when a critical film thickness is reached.

Others have studied the influence of friction-generated contact temperatures on the stress field. Stress analysis of a semi-infinite body due to a moving heat source has been investigated by Burton and Tseng [22] for the two dimensional plane stress case. They found that the thermal compressive stress in the body is approximately ten times that from the normal load. Huang and Ju [23] studied the thermo-mechanical cracking problem of a semi-infinite elastic medium due to an asperity travelling at high speed. In their analysis, the isothermal and thermal stress fields in a semi-infinite body induced by a moving contact region are calculated. The combined stress field is then used to investigate the surface cracking problem.

In the proposed thermomechanical wear theory, the influence of the contact temperature on wear will be studied, and wear is assumed to result from the yielding of materials. The yielding condition is governed by the temperature dependent yield strength and the stress field of the material. The stress field is combined from the isothermal stress field generated by surface tractions and the thermal stress field induced by the frictional heating. The theory will focus on the contacting asperity which is stationary relative to the surface loadings because higher temperatures and more prominent thermal effects are expected for that case.

Experiments are being conducted to provide the measurement of friction and the size of the real area of contact to be used in the theory to calculate wear rate. Measurements of the actual wear rate which is to be compared with the calculated wear rate will also be performed. The appropriateness of the theory and the answer to the question of whether or not the frictional heating is an important factor in the wear process will be determined by the comparison of measured and predicted wear rates.

PROPOSED THEORY

I. Thermomechanical model:

Suppose asperities in a tribo-contact are spherical, the area of contact is circular and the pressure distribution is hemispherical. If the coefficient of friction is constant over the contact region, there are three similarly distributed surface loadings on the real area of contact, frictional heat, tangential and normal surface tractions. When the contacting asperities are sparsely distributed over the apparent area of contact, the analysis of a single asperity contact can be taken as loadings on a semi-infinite body. Fig. 1 shows the loadings and the reference coordinates of this model.

Surface loadings are divided into a traction free thermal loading and an isothermal traction loading. The temperature distribution will first be deduced from the thermal analysis of the frictional heating, and the corresponding thermal stress field will be obtained from quasistatic thermoelasticity analysis. The isothermal stresses will be calculated from the available solutions. The isothermal and thermal stress fields are superimposed to give the stress field to determine yielding.

II. Temperature distribution:

Suppose the thermal energy generated by friction at the contact is the only heat source and to be transferred through the circular contact into a semi-infinite body. If the origin of the space coordinates is chosen at the center of the circular contact area, (refer to Fig. 1), the transient temperature field is axisymmetric and the governing equation can be formulated as

$$\frac{\partial^2 T}{\partial \rho^2} + \frac{1}{\rho}\frac{\partial T}{\partial \rho} + \frac{\partial^2 T}{\partial Z^2} = \frac{a^2}{D}\frac{\partial T}{\partial t}$$

where T is the temperature rise above the bulk temperature, "a" is the contact area radius, and D is the thermal diffusivity.

If the surface outside the frictional contact is insulated, the boundary conditions are

$$\frac{\partial T}{\partial \rho} = 0 \quad \text{at} \quad \rho = 0 \qquad\qquad \frac{\partial T}{\partial Z} = \begin{cases} -\dfrac{Q_o a}{k}\sqrt{1-\rho^2} & 0 \le \rho \le 1 \\ 0 & \rho > 1 \end{cases}$$

$$T \to 0 \quad \text{as} \quad \rho \to \infty \quad \text{and} \quad Z \to \infty$$

where Q_0 is the peak frictional heat input rate per unit area to the surface ($Q_0 = \gamma f P_0 V$), γ is the heat partition function and k is the thermal conductivity. And the initial condition is

$$T = 0 \quad \text{at} \quad t = 0$$

The problem will be solved by applying the Hankel zero order transform with respect to ρ and the Laplace transform with respect to time [24].

III. Stress analysis:

a) Thermal stress analysis:

Consider a linear thermoelastic, homogeneous, constant properties solid, in which the temperature distribution is the solution discussed in (II) above, the thermal stresses due to this temperature field can be obtained from thermoelasticity. Using the displacement formulation or the Navier's Equation, with body force equal to zero, the equation can be written as

$$(\lambda + \mu)u_{k,ki} + \mu u_{i,jj} - \beta T_{,i} = 0$$

where u_i is the displacement in the i direction, λ and μ are Lame's elastic constants, $\beta = \frac{E\alpha}{1-\nu}$, α is the linear thermal expansion coefficient, and E is the elastic modulus.

Introducing the thermoelastic displacement potential, ϕ, $\phi_{,i} = u_i$, and substituting to the above equation. Integration and equating the integration function to zero gives

$$\nabla^2\phi = -\frac{\alpha(1+\nu)}{(1-\nu)} T$$

The boundary conditions are

$$\frac{\partial \phi}{\partial \rho} = 0 \quad \text{at} \quad \rho = 0 \;, \qquad \phi \to 0 \quad \text{as} \quad \rho \to \infty$$

$$\frac{\partial \phi}{\partial Z} \to 0 \quad \text{as} \quad Z \to \infty \;, \qquad \frac{\partial^2 \phi}{\partial Z^2} - \nabla^2\phi = 0 \quad \text{at} \quad Z = 0$$

The double integral transform techniques used in solving the temperature distribution will also be used to solve this equation. The thermal stress field can be obtained from ϕ by the following relations

$$\sigma_{r_1} = 2\mu\left\{\frac{\partial^2\phi}{\partial r^2} - \nabla^2\phi\right\}, \qquad \sigma_{\theta_1} = 2\mu\left\{\frac{1}{r}\frac{\partial\phi}{\partial r} - \nabla^2\phi\right\}$$

$$\sigma_{z_1} = 2\mu\left\{\frac{\partial^2\phi}{\partial z^2} - \nabla^2\phi\right\} \quad \text{and} \quad \sigma_{rz_1} = 2\mu\,\frac{\partial^2\phi}{\partial r\partial z}$$

The thermal stresses determined from the above equations usually cannot completely satisfy the traction boundary conditions [25,26]. Therefore, a fictitious traction applied to the boundary is needed to suppress the unsatisfied boundary condition. The stress field induced by this traction is superimposed to that determined from ϕ to complete the thermal stress field.

In the axisymmetric case, it is convenient to use Love's function, L [27]. The stress field generated by surface tractions can be found from the following equations:

$$\sigma_{r_2} = \frac{\partial}{\partial z}\left[\nu\nabla^2 L - \frac{\partial^2 L}{\partial r^2}\right], \qquad \sigma_{\theta_2} = \frac{\partial}{\partial z}\left[\nu\nabla^2 L - \frac{1}{r}\frac{\partial L}{\partial r}\right]$$

$$\sigma_{z_2} = \frac{\partial}{\partial z}\left[(2-\nu)\nabla^2 L - \frac{\partial^2 L}{\partial z^2}\right] \quad \text{and} \quad \sigma_{rz_2} = \frac{\partial}{\partial r}\left[(1-\nu)\nabla^2 L - \frac{\partial^2 L}{\partial z^2}\right]$$

where L is the solution of the biharmonic equation $\nabla^4 L = 0$. In the axisymmetric case, the solutions are available in different forms. Because there are only two traction boundary conditions, σ_{rz} and σ_z, we can choose a linear combination of two kinds of the available solutions with two undetermined constants as the solution, and determine the constants by satisfying the following traction boundary conditions

$$\sigma_z = \sigma_{z_1} + \sigma_{z_2} = 0 \quad \text{and} \quad \sigma_{rz} = \sigma_{rz_1} + \sigma_{rz_2} = 0 \quad \text{at} \quad z = 0$$

The thermal stress field will be completed by superimposing the stress fields obtained from ϕ and L.

b) Isothermal stress analysis:

The basic isothermal stress field of an elastic semi-infinite body subjected to surface loading has been determined by Boussinesq [28], Flamant [29] and Cerruti [30]. A complete isothermal stress field in a semi-infinite body due to hemispherical-distributed normal tangential surface tractions has been obtained by Hamilton and Goodman [31]. The stress components are expressed in algebraic equations containing complex variables. Computation of stresses based on these equations have been carried out. The constant contours of the ratio of the square root of the second invariant of the stress deviator tensor to the peak Hertzian pressure is shown in Fig. 2. This solution will be adopted as the isothermal stress field due to surface tractions.

IV. Material response:

Mechanical and thermal properties of materials are temperature dependent. When the analysis covers a wide range of temperature, the change of material properties with temperature should be considered. However, the analysis incorporating the temperature dependent properties is mathematically complex and beyond the scope of this research. Therefore, constant property analysis is used to determine the temperature and the stress fields, and the material responses to the temperature variation will be used in obtaining the temperature dependent yielding conditions.

Generally, material yield strength decreases with increasing temperature at fixed strain rate. A linearly or exponentially decreasing function of temperature will be assumed for the yielding strength [32,33]. The maximum distortion energy yielding criterion will be used with this temperature-dependent yielding strength to determine the inelastic deformation.

EXPERIMENT

The experiments will be performed in a pin-on-disk tribometer with unlubricated ceramic pins and single crystal sapphire disk. The experimental system is shown schematically in Fig. 3. The wear rate is measured by using a linear voltage displacement transducer attached to the disk spindle and detecting the vertical displacement of the disk. The friction is measured with a load cell installed in the supporting truss. These two measurements are recorded with the digital storage oscilloscope.

The radius of the contact region will be measured by an infrared scanning system. The IR system detects the thermal radiation from the surface and creates a thermograph, which can be recorded with a video recorder. The grey level of each pixel in an IR thermograph represents a temperature level. Using a digital video processor, the grey levels are converted into numerical values and sent to a computer. The calibration of the numerical value to temperature provides the contact temperature measurement. When the measured surface temperature is compared with that calculated at $r = a$ from the theory, a theoretical contact radius can be obtained. With friction and contact radius being measured, the theoretical wear rate can be calculated and compared to that measured.

SUMMARY

A thermomechanical wear theory is proposed which includes the thermal effect due to the frictional heating. It is used to evaluate the state of the material at the tribo-contact and to determine a region where material is inelastically deformed. When this region adjoins to the surface, material flows and wear takes place. Wear rate is theoretically deduced by calculating the volume of this inelastic region. The experimental work will provide a measure of the adequacy of the theory by comparing the wear rates.

Good agreement of the predicted and measured wear rates will imply that thermomechanical wear is the dominant wear mechanism. Even if the theory fails to give a good estimation, the ratio of the theoretical wear rate to that measured will indicate the relative importance of the thermomechanical wear. This ratio may be dependent on the operating conditions. Therefore, the proposed study provides a thermomechanical wear theory which will either give a good wear rate prediction or identify a tribological operating range in which the thermomechanical wear is dominant. When the dominant wear mechanism is determined in a specific tribological operation, wear can be controlled by varying the operating conditions.

It is expected that the thermomechanical wear theory will provide information about how wear should be treated in an engineering design involving frictional heating, and indicate the dominant factors which can be controlled to reduce wear.

REFERENCES

1. "Survey of Possible Wear Mechanisms," J. T. Burwell, *WEAR*, *1*, 119-141 (1957).
2. "Wear Theory and Mechanisms," J. F. Archard, *Wear Control Handbook*, M. B. Peterson and W. O. Winer, editors, American Society of Mechanical Engineers, New York, 35-80 (1980).
3. "Theoretical Study of Temperature Rise at Surfaces of Actual Contact Under Oiliness Lubricating Conditions," H. Blok, *Proceedings General Discussion on Lubrication and Lubricants, London, 2*, Institution of Mechanical Engineers, London, 222 (1937).
4. "The Temperature of Rubbing Surfaces," J. F. Archard, *WEAR*, *2*, 438-455 (1958-1959).
5. "The Influence of Thermal Expansion on the Friction and Wear Process," J. R. Barber, *WEAR*, *10*, 155-159 (1967).
6. "Thermoelastic Instabilities in the Sliding of Conforming Solids," J. R. Barber, *Proceedings of the Royal Society, London, A312*, 177-188 (1969).
7. "Distortion of the Semi-Infinite Solid Due to Transient Surface Heating," J. R. Barber, *International Journal of Mechanics and Science*, *14*, 377-393 (1971).
8. "Some Thermoelastic Contact Problems Involving Frictional Heating," J. R. Barber, *Quarterly Journal of Mechanics and Applied Mathematics*, *29*, 1-13 (1976).
9. "The Transient Thermoelastic Contact of a Sphere Sliding on a Plane," J. R. Barber, *WEAR*, *59*, 21-29 (1980).
10. "Thermoelastic Displacements and Stresses Due to a Heat Source Moving Over the Surface of a Half-Plane," J. R. Barber, *Journal of Applied Mechanics*, *Transactions of the American Society of Mechanical Engineers*, *51*, 636-640 (1984).
11. "Effect of Initial Surface Curvature on Frictionally Excited Thermoelastic Phenomena," R. A. Burton, *WEAR*, *27*, 195-207 (1974).
12. "Large Disturbance Solution for Initially Flat Frictionally Heated Thermoelastically Deformed Surfaces," R. A. Burton and V. Nerlikar, *Journal of Lubrication Technology*, *Transactions of the American Society of Mechanical Engineers*, *97*, 546-551 (1975).

13. "Thermal Deformation in Frictionally Heated Contact," R. A. Burton, *WEAR*, *59*, 1-20 (1980).
14. "Thermal Aspect of Galling of Dry Metallic Surface in Sliding Contact," F. F. Ling and E. Saibel, *WEAR*, *1*, 80-91 (1957).
15. "Probable Interface Temperature of Solids in Sliding Contact," F. F. Ling and S. L. Pu, *WEAR*, *7*, 23-24 (1964).
16. "On Temperature Transients in Sliding Interface," F. F. Ling, *Journal of Lubrication Technology*, *Transactions of the American Society of Mechanical Engineers*, *91*, 397-405 (1969).
17. "A Thermal, Thermoelastic and Wear Simulation of a High Energy Sliding Contact Problem," F. E. Kennedy and F. F. Ling, *Journal of Lubrication Technology*, *Transactions of the American Society of Mechanical Engineers*, *96*, 497-507 (1974).
18. "The Effect of 'Hot-Spot' Temperatures on the Unlubricated Wear of Steel," T. F. J. Quinn, *American Society of Lubrication Engineers Transactions*, *10*, 158-168 (1967).
19. "Frictional Heating and the Oxidational Theory of Wear," D. M. Rowson and T. F. J. Quinn, *Journal of Physics D: Applied Physics*, *13*, 209-219 (1980).
20. "Review of Oxidational Wear, Part I: The Origins of Oxidational Wear," T. F. J. Quinn, *Tribology International*, *16*, 257-271 (1983).
21. "Review of Oxidational Wear, Part II: Recent Developments and Future Trends in Oxidational Wear Research," T. F. J. Quinn, *Tribology International*, *16*, 305-315 (1983).
22. "Thermal Stress in a Two-Dimensional (Plane Stress) Half-Space for a Moving Heat Input," R. A. Burton, *WEAR*, *79*, 1-9 (1982).
23. "Thermomechanical Cracking Due to Moving Frictional Loads," J. H. Huang and F. D. Ju, *WEAR*, *102*, 81-104 (1985).
24. "Temperature Prediction in Mechanical Components, an Analytical Approach," B. Gecim, Ph.D. Thesis, School of Mechanical Engineering, Georgia Institute of Technology (1984).
25. *Thermoelasticity*, H. Parkus, Springer-Verlag/Wien, New York, 18 (1976).
26. *Theory of thermoelasticity with applications*, J. L. Nowinski, Sijthoff & Noordhoff International Publishers, Alphen aan den Rijn, The Netherlands, 265 (1978).
27. *A Treatise on the Mathematical Theory of Elasticity*, A. E. H. Love, Dover, New York, 274 (1944).
28. *Application des potentials a l'etude de l'equilibre et du mouvement des solides elastiques*, J. Boussinesq, Gauthier-Villars, Paris, France (1885).
29. "Sur la lepartition des pressions dans un solide rectangulaire charge transversalement," M. Flament, *Comptes Rendus*, *114*, 1465-1468 (1892).
30. "Ricerche intorno all'equilibrio de corpi elastici isotropi," V. Cerruti, *Atliacademi nazl. Lincei, Mem. classe sci. fis., mat. c nat.*, *13*, 81 (1892).
31. "The Stress Field Created by a Circular Sliding Contact," G. M. Hamilton and L. E. Goodman, *Journal of Applied Mechanics*, *33*, 371-376 (1966).
32. "Thermal Stress in a Viscoelastic-Plastic Plate with Temperature-Dependent Yield Stress," H. G. Landau, J. H. Weiner and E. E. Zwicky, *Journal of Applied Mechanics*, *27*, 297-302 (1960).
33. *Introduction to Ceramics*, W. D. Kingery, H. K. Bowen and D. R. Uhlmann, 2nd edition, John Wiley and Sons, New York (1976).

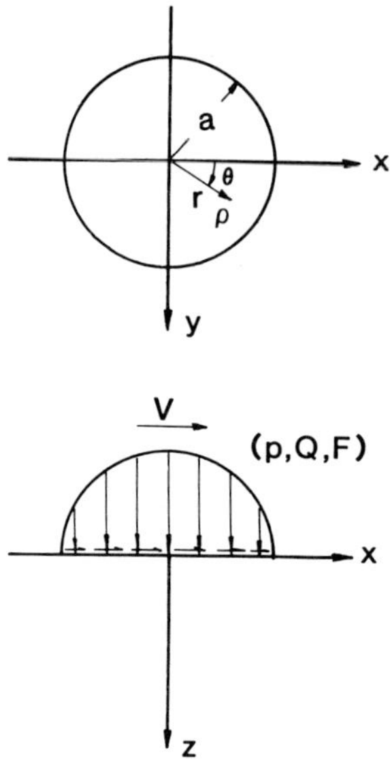

$$\frac{p \text{ (normal pressure)}}{P_0 \text{ (peak Hertzian pressure)}} = \sqrt{1-\rho^2}$$

F (tangential pressure) $= fp$

f: coef. of friction

$$\frac{F}{f P_0} = \sqrt{1-\rho^2}$$

$$\frac{Q \text{ (heat flux input rate)}}{Q_0 \text{ (peak heat flux input rate)}} = \sqrt{1-\rho^2}$$

$$\rho = r/a$$

Fig. 1. Surface loadings and the reference coordinates in the thermomechanical model.

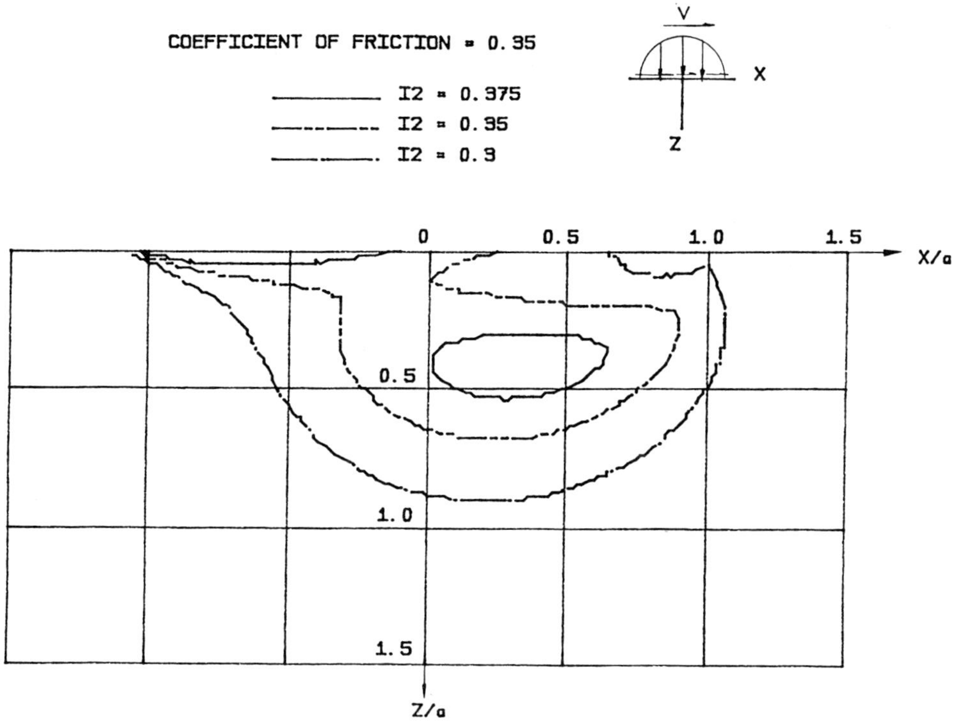

Fig. 2. Constant contours of constant I_2 at y = 0 plane.

$I_2 = \sqrt{\text{(second invariant of stress tensor)}} / P_0$.

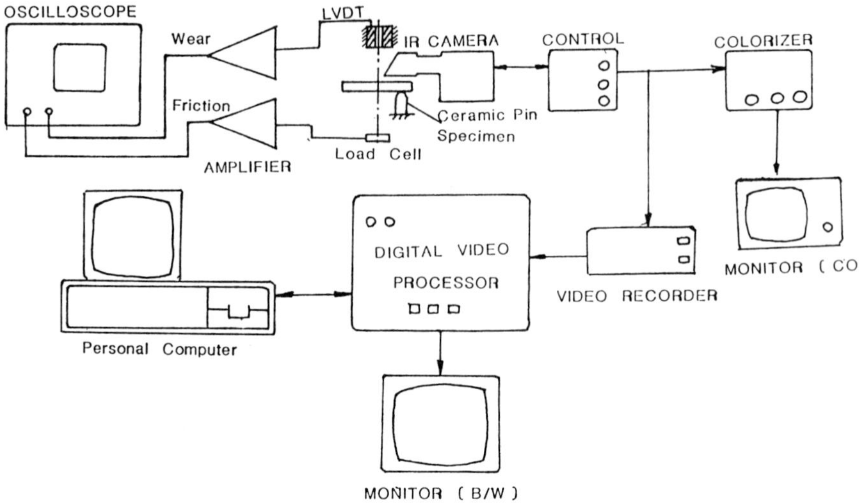

Fig. 3. Schematic illustration of the experiment equipment and the installatio

4.14 PREDICTIVE MODELS FOR SLIDING WEAR

Said Jahanmir

INTRODUCTION

Wear is defined as "damage to a solid surface, generally involving progressive loss of material due to relative motion between that surface and a contacting substance or substances" [1]. Examination of worn machine elements indicates that the wear process is rather complex and can occur by various mechanisms [2]. Wear can be regarded as the result of the surface being stressed mechanically, thermally, chemically, or electrically. These processes may occur independently or simultaneously to cause material removal. When two surfaces are rubbed together, in the absence of any foreign abrasive particles, the wear process is classified as "sliding wear." In sliding wear, various processes can operate to generate wear particles.

Professors Rigney and Godet have presented an excellent review of two aspects of unlubricated wear; mainly the role of transfer layers, and loose wear debris or third-body. However, there are other aspects of sliding wear that should be considered in modeling of wear. Since wear occurs at the microcontacts, the chemical and physical state of surfaces play a dominant role on the wear process. The effect of absorbed surface layers and tribo-chemical reactions are discussed in this paper; and a few guidelines are presented for modeling of sliding wear. Some of the ideas presented here were discussed at a recent symposium [3].

THE MECHANISMS AND PROCESSES OF WEAR

Over the past years, many wear mechanisms and wear processes such as adhesion, abrasion, delamination, fatigue, and corrosion, have been introduced. Most of these mechanisms are supported through laboratory test results. It is, however, unfortunate that the range of applicability of each wear process has not been defined. The wear process in most tribological applications involves more than one mechanism. This complicates the identification of the wear mechanism and selection of an appropriate solution to combat wear. It is, therefore, desirable to identify and classify the possible wear mechanisms in important tribological applications. In such studies, the link between the fundamental wear mechanisms and the actual wear process needs to be clearly identified. Special attention should be given to the wear transitions that are sometimes critical in determining the overall life of a system.

Perhaps the best form for classification of wear mechanisms is through an organized form such as wear maps [4]. These wear maps should include the range of effective operating conditions for each wear mechanism, (such as load, temperature, speed, etc.) and the specific wear rate data in the form of wear coefficients. The wear maps should be developed for various engineering materials, and could be computerized into data bases for easy access. Wear maps should also be developed for tribological applications, showing the possible wear processes and limiting range of each mechanism for different applications. For example, a set of wear maps for cam and lifter contact of an IC engine would consist of wear mechanisms and the effect of system operating variables, lubricant chemistry, and material combination.

The majority of wear particles generated in a sliding contact, in a dry or a lubricated environment, are laminar in shape [5]. These particles can be generated by asperity deformation, plowing, delamination or adhesion. The controlling parameters such as load, speed, temperature, and environment strongly influence the process of wear particle formation. Under very light loading situations, wear primarily occurs by deformation and fracture of asperities. As a result of material transfer to the opposite surface, or at a more moderate loading situation, the wear process changes to plowing. Under conditions where the loads are larger and the sliding condition is more severe, wear particle formation occurs by accumulation of plastic deformation causing delamination. Under severe sliding conditions, where the adhesive forces at the interface are strong, wear can occur by adhesion and material transfer. The operating parameters of a tribo-system determines the controlling wear process. It has been shown, for example, that the wear processes can be ranked in the following order; asperity deformation, plowing, delamination, and adhesion, by increasing either the friction coefficient [6] or the applied normal load [7]. Transition from one wear process to another is also dependent on the environment and formation of tribochemical layers.

The coefficient of friction is sensitive to any physical or chemical changes at the interface between two sliding surfaces. It is well-recognized that adsorption of species from the environment or the lubricant generally decreases friction, through reduction of metal-metal contact. As an example, long-chain polar compounds added to lubricating oils decreases friction through adsorption and reduction of adhesion tendencies between the contacting surfaces [8,9,10]. This reduction is related to the molecular structure of the compounds and the nature of the interaction with the surfaces [9]. The reduction in friction is often accompanied by reduction in metal-metal adhesion and reduced wear [6]. The decreased adhesion would tend to reduce or eliminate the possibility of transfer layer formation.

The details of wear process are dependent on the chemical composition, structure and properties (both physical and chemical) of the surfaces in contact. It is recognized that tribo-chemical reactions occur at a much faster rate at the interface as compared to static reactions. The exact mechanism for this enhancement has not been identified, but several suggestions have been made. These include exposure of fresh surface, exo-electron emission, surface defect formation and increased temperature. The most probable mechanisms by which friction increases the reaction rate are the removal or thinning of the reaction layer by wear, or the increase in diffusion of reagents by a combination of temperature increase, and introduction of diffusion paths through plastic deformation. The nature of tribo-chemical reactions and kinetics of surface film formation need to be determined for specific situations where the films have controlling effects on friction and wear. Such studies are important in the understanding of wear rate reduction by antiwear additives, and the general effect of environment on wear.

It is imperative to characterize the contact surfaces and to determine whether the wear particles originate from a tribo-layer or from the surface with the same composition and structure as the bulk material. It should be pointed out that in many cases, formation and breakdown of surface layers control the rate of wear. Special attention should be given to the thermal and environmental effects which are conducive to tribo-chemical reactions and surface layer formation. It is also important to investigate the effect of wear particles on the wear process. It should be determined whether the wear particles are re-deposited to form "mechanically alloyed" surface layers or act as abrasive particles.

The wear mechanisms that have been advanced in the past mostly deal with wear coefficients that are larger than 10^{-6}. However, the drive toward higher reliability and extended life in many tribological applications require smaller wear coefficients. In this regard, special attention should be given to identifying the wear mechanisms that take place at microscopic levels, such as lubricated contacts and magnegic recording devices.

MODELING OF WEAR

There has been much progress in modeling of fluid film lubrication phenomena such as hydrodynamic, elastohydrodynamic, and gas film lubrication [11]. But there is little progress toward general models for friction, wear, and boundary lubrication. The primary reason for lack of predictive models in these areas is the complexity of the interactions at interfaces which require interdisciplinary approaches to the problems. The models should be based on known physical principles and should predict the performance of typical tribological contacts. Such models should have the capability of being used in design and material selection, and should predict the value of friction coefficient, wear rates, and the maximum life of the components.

At least two approaches can be taken in developing models. One approach is based on empirical data collection which to some extent is currently being done, although often not in a coordinated manner. This approach involves characterization of surfaces, collection of tribological data in laboratory tests, correlation of data with field tests, and sorting out the data to build an empirical model. This approach is quite time-consuming and must be coordinated to be successful. A second approach involves generation of models based on known physical principles, followed by confirmation through data collection. The model is then modified to agree with experimental facts. Obviously, there are advantages and disadvantages for each approach. Perhaps the best method is a combination of both approaches.

The success of fluid film lubrication models is essentially due to the advances in continuum analysis of fluid mechanics, which can ignore the molecular details of the lubrication phenomena. However, any useful models for friction, wear, and boundary lubrication must incorporate the physical and chemical processes that occur at the interface. Furthermore, the details of surface film formation and breakdown must be included in the models. This is an extremely difficult task and can only be approached as a well-planned interdisciplinary research. A possible stepwise approach may be to first organize and evaluate the present published information, followed by identification of information gaps. The next step would be to plan a comprehensive team approach such that molecular models, chemical reactions, film formation, and continuum mechanics models proceed as a joint collaborative effort between experts in each area. The theoretical models must be checked and guided by experimental data to assure the general applicability of the models.

REFERENCES

1. "Standard Terminology Relating to Erosion and Wear," ASTM CR 40-82, *Annual Book of ASTM Standards*, *3.02*, American Society for Testing and Materials (1983).
2. "Wear of Machine Elements," F. T. Barwell, *Fundamentals of Tribology*, N. P. Suh and N. Saka, editors, MIT Press, Cambridge, MA, 401-442 (1980).
3. "Future Directions in Tribology Research," S. Jahanmir, submitted for publication to *Journal of Tribology*, *Transactions of the American Society of Mechanical Engineers*.
4. "Wear Mechanisms Maps," S. C. Lim and M. F. Ashby, *Acta Metallurgica*, *35*, 1-24 (1985).
5. "On Mechanics and Mechanisms of Laminar Wear Particle Formation," S. Jahanmir, *Advances in the Mechanics and Physics of Surfaces*, R. M. Latanision and T. E. Fischer, editors, *3*, Harwood, New York, NY, 261-332 (1986).
6. "The Relationship of Tangential Stress to Wear Particle Formation Mechanisms," S. Jahanmir, *WEAR*, *103*, 233-252 (1985).
7. "Transition of Lubricated Wear of Carbon Steel," Y. Iwai and K. Endo, *Wear of Materials*, S. K. Rhee, A. W. Ruff and K. C. Ludema, editors, American Society of Mechanical Engineers, New York, NY, 656-662 (1981).
8. "An Adsorption Model for Friction in Boundary Lubrication," S. Jahanmir and M. Beltzer, *American Society of Lubrication Engineers Transactions*, *29*, 423-430 (1986).
9. "Effect of Additive Molecular Structure on Friction and Adsorption," S. Jahanmir and M. Beltzer, *Journal of Tribology*, *Transactions of the American Society of Mechanical Engineers*, *108*, 109-116 (1986).
10. "Role of Dispersion Interactions Between Hydrocarbon Chains in Boundary Lubrication," M. Beltzer and S. Jahanmir, *American Society of Lubrication Engineers Transactions*, *30*, 47-54 (1986).
11. "The Reynolds Centennial. A Brief History of the Theory of Hydrodynamic Lubrication," O. Pinkus, *Journal of Tribology*, *Transactions of the American Society of Mechanical Engineers*, *109*, 1-8 (1987).

4.15 SURFACE DEFORMATION CONSIDERATIONS FOR ROLLING WITH INCIPIENT SLIDING

Lavern D. Wedeven

Attention is focused on the material's response to stress (normal and tangential) in connection with the modelling of material removal by wear as well as the resulting frictional force to produce sliding. Tribological models are confined generally to relative motion in pure sliding where the surface material of one body is in continual contact with the passing of the surface material of another body. The case of combined rolling and sliding (particularly incipient sliding), while important from a practical perspective, may also be very useful for tribological model development.

Two rolling bodies can depart from a "free rolling" condition if they are called upon to transmit a tangential force (T). If the tangential force is less than the limiting friction force ($T < \mu P$, where μ and P are friction coefficient and normal load respectively) the two bodies will not give rise to gross sliding but to creep which results in a contact that is divided into "stick" and "slip" regions (see Fig. 1). The strain in the stick region is dominated by elastic properties. Also, the difference between the tangential strains in the two bodies in the stick region leads to an apparent slip where in one revolution the distance covered is slightly greater than the undeformed perimeter of a circular body. The strain in the "slip" region is controlled by the local coefficient of friction and normal load. The size of the "stick" and "slip" regions is a function of the tangential force (T). A small tangential force causes micro-slip at the trailing edge of the contact. As the tangential force increases, the slip region extends forward until when $T = \mu P$ the slip region reaches the leading edge of the contact and gross sliding occurs. A typical distribution of tractive stress in the stick and slip regions is shown in Fig. 1(b) and the effect of tractive force on slip (creep curve or traction curve) is shown in Fig. 2 (see, e.g., Ref. 1 for more detail).

The initial slope of the traction curve in Fig. 2 reflects the elastic response (shear modulus) of the material in the contact region. For "clean" interfacial conditions, the strain affected zone extends below the surface (near-surface region) and the initial slope reflects the shear modulus of the bulk materials. For other than "clean" conditions (e.g., liquid or solid films, oxides, compacted islands of oxides, wear debris) the stiffness of the "third body material" will contribute to a portion of the actual initial small strain response. For larger strains, of course, the shear properties of the "third body material" will control the resulting tractive stress. Thus, it is possible to use the initial slope of a traction vs. slip curve to obtain the shear modulus of the strain affected zone [2] or even an effective viscosity of a solid film [3,4]. From Ref. 2 an effective shear modulus (G) can be obtained from

$$G = \pi m h p_o/(4a)$$

where m = initial slope of traction curve

h = depth of strain affected zone

p_o = maximum Hertz pressure

a = contact half-width in rolling direction

For high strain conditions (gross sliding throughout the contact) the maximum tractive force as shown in Fig. 2 is determined by the shear strength of the "third body material" which is a function of the interfacial conditions within the contact (e.g., pressure, temperature, shear rate) as well as the composition and local distribution of the "third body material" at a given instant in time.

The entire traction curve then, from initial slip to maximum tractive force, reflects deformation properties arising from a zone below the surface to the shear plane which gives rise to the frictional response. Conceptually, empirical surface deformation properties can be derived from carefully obtained traction measurements from incipient to gross sliding. These derived properties can then be utilized in conjunction with the tribological modelling of constitutive parts of the total problem so that the over-all modelling effort can be related to some degree of reality.

TRACTION EXPERIMENTS

An example of some recent traction experiments reflects some of the features discussed above. Traction vs. slip measurements were obtained under rolling conditions with material specimen and test configuration shown in Fig. 3. Fig. 4 shows the unlubricated traction curves for two material pairs of widely differing elastic properties. The upper curve, showing a large initial slope, was obtained with high elastic modulus materials (silicon nitride) and the lower curves, showing a lower initial slope, were obtained with lower elastic modulus materials (bulk material BG42 stainless steel). The tests were performed under constant contact pressure (2068 MPa) and the traction is plotted as traction coefficient (T/P).

The four traction curves show widely differing maximum traction coefficient as a result of differing shear properties at the interface. The unlubricated silicon nitride pair shows a maximum traction coefficient larger than 1.0. The three curves for the BG42 stainless steel pairs start off with the same initial slope but reach a traction level characteristic of their surface treatment. The upper of the three curves was obtained with salt bath nitrided (SBN) surfaces, the middle curve was obtained with SBN plus an oxidizing (O) treatment and the lower curve was obtained with an untreated BG42 ball against a SBN plus O treated disk. The lower traction level reached with the latter two curves is obviously due to the lower shear strength of the oxidized surfaces. The transition in the curve for the untreated BG42 ball against the oxidized disk is due to depletion of the oxidized layer (or adsorbed films) on the disk. It is conceivable that by operating at progressively increasing slip levels from pure rolling an estimate of the shear strength of surface films could be obtained.

In the case of rolling with incipient sliding there is no relative sliding motion between the contacting bodies at the interface in the stick region. The stick region can encounter high traction stresses, but most of the tribological interactions and energy dissipation are likely to be confined to the slip region. The close connection between traction measurements and detailed examination of the surfaces would seem to be essential for tribolgical model development.

Some typical surface features are shown in Fig. 5 and Fig. 6. Fig. 5 shows the rolling contact tracks of the SBN+O treated BG42 ball and disk specimens following the traction test shown in Fig. 4. In this case the shear due to slip is accommodated by the deformation and removal of material in the oxygen rich layer (perhaps a few microns thick). Fig. 6 shows the rolling tracks of a silicon nitride ball and AISI 440C stainless steel disk following a traction test (not shown in Fig. 4). In this case the tribological response to tractive stress and slip results in apparent oxide film formation on the stainless steel surface and transverse stress cracks in the silicon nitride.

It would appear that the case of rolling with incipient sliding could benefit the adaptation of surface deformation models to predict tribological behavior. Tribological modelling with closely connected theory and experiment is an obvious requirement. Less obvious is the need for modelling which enlightens pathways for tribological improvement in addition to prediction of observed behavior.

REFERENCES

1. *Contact Mechanics*, K. L. Johnson, Cambridge University Press, Cambridge (1985).
2. "Rolling, Slip and Endurance Traction Measurements on Low Modulus Materials," J. L. Tevaarwerk, *NASA CR 174909* (July 1985).
3. "The Effect of Solid Film Lubricants on the Stability of Rolling Element Bearings," J. W. Kannel and T. L. Merriman, *ASME Paper No. 86-Trib-61* (1986).
4. "Traction Characteristics of Graphite Lubricants at High Temperature," R. A. Pallini and L. D. Wedeven, submitted for the 42nd Annual ASLE Conference, Anaheim, CA, May 11-14, 1987.

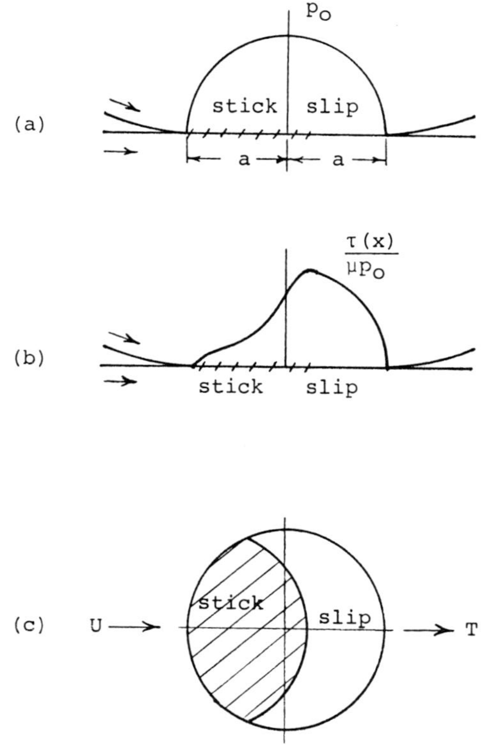

Fig. 1. **Rolling with incipient sliding for circular contact (T < μP).**

(a) normal pressure distribution
(b) tractive stress distribution
(c) expected stick and slip regions

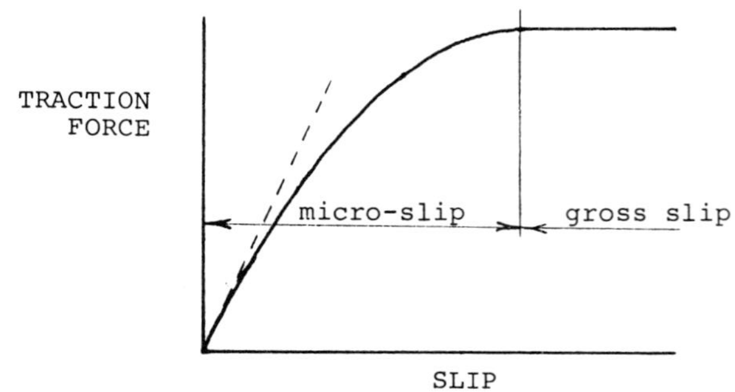

Fig. 2. **Traction vs. slip curve (creep) for unlubricated surfaces.**

Fig. 3. Traction test rig.

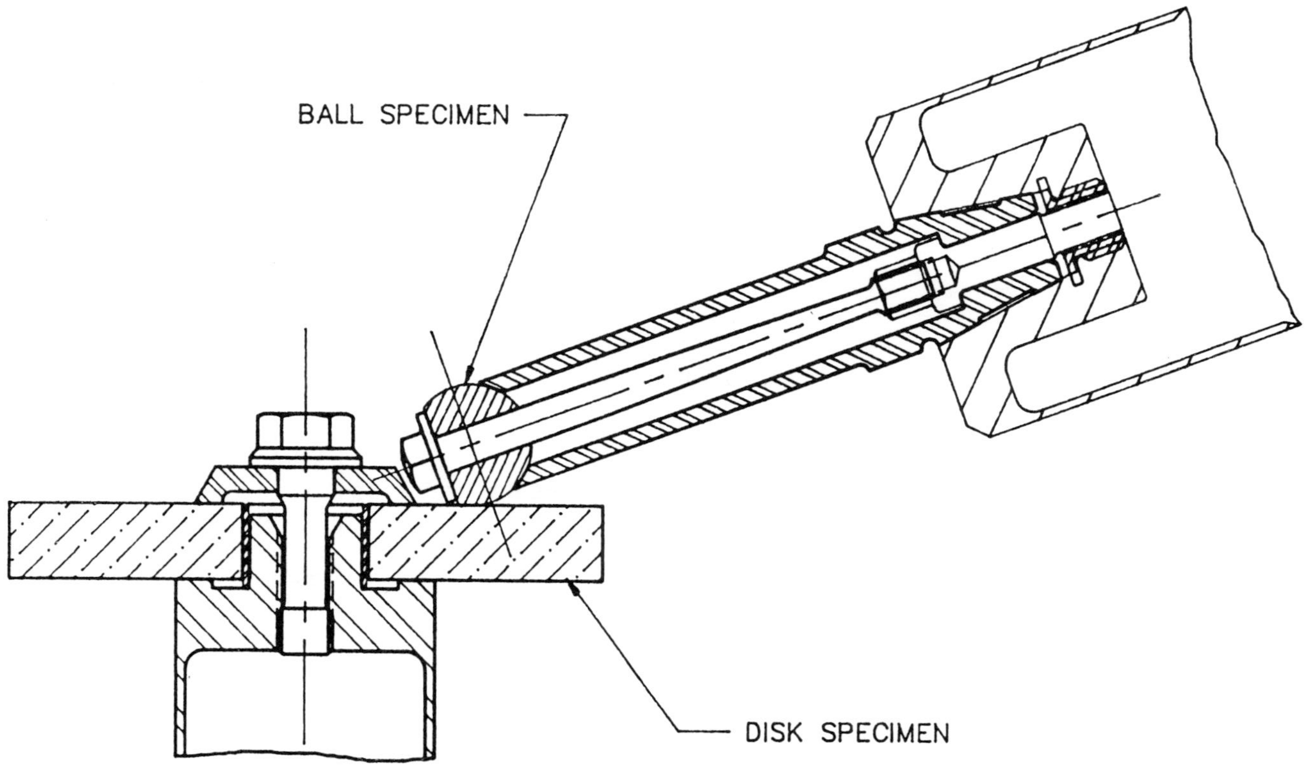

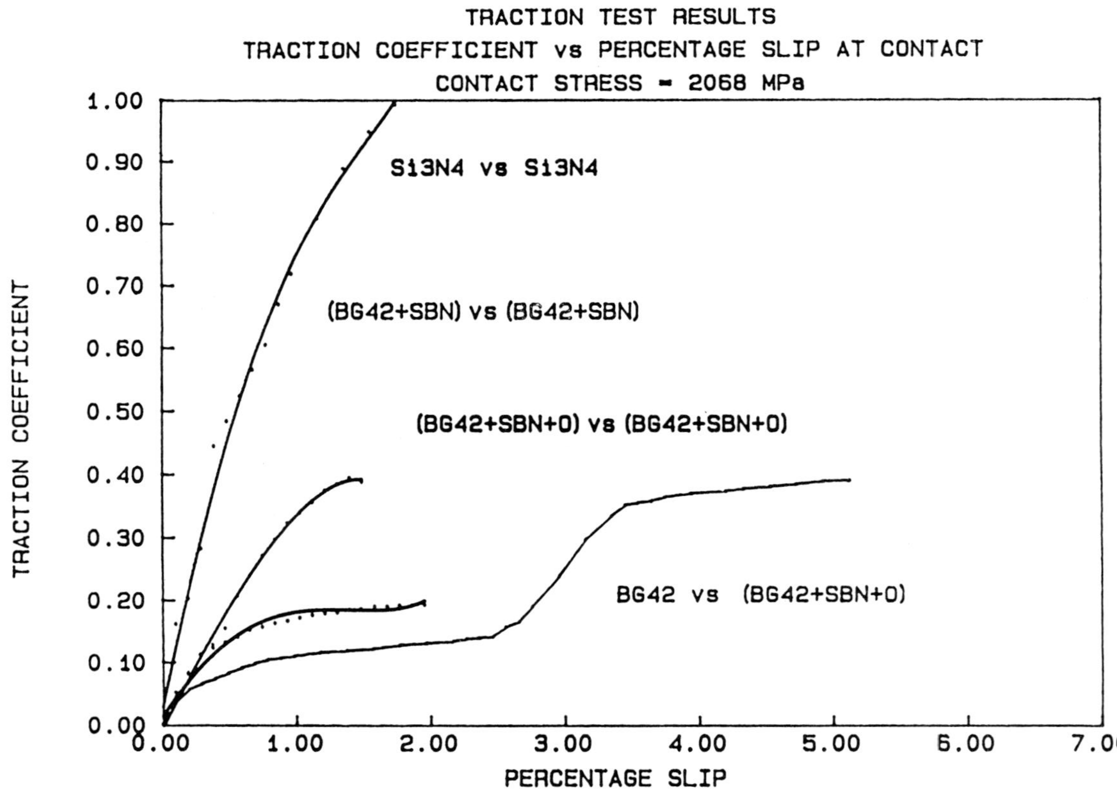

Fig. 4. Traction curves for unlubricated silicon nitride and unlubricated stainless steel (BG42) with combinations of salt bath nitriding (SBN) and oxidizing (O) treatments. Rolling speed 3.8 m/s, room temperature.

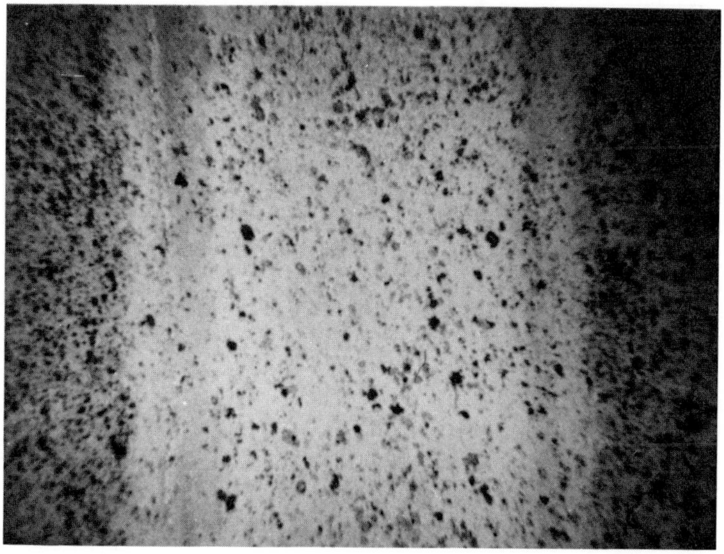

A

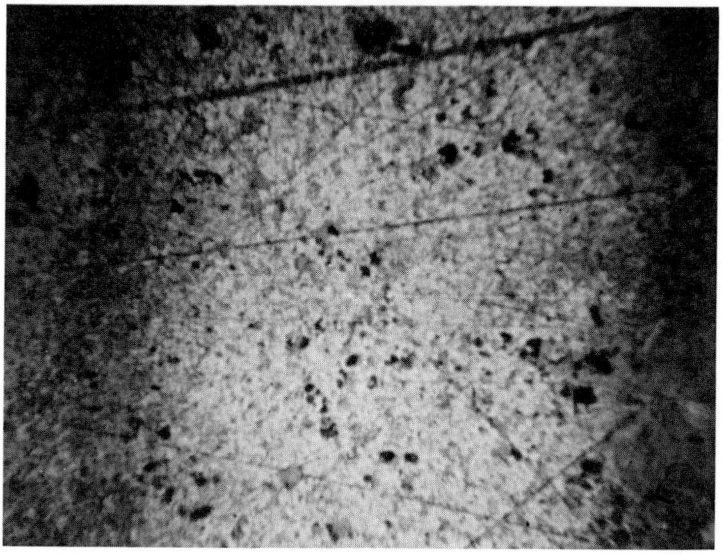

B

Fig. 5. Photomicrographs of rolling tracks following traction test with BG42 stainless steel specimens processed with salt bath nitriding plus oxidizing surface treatments; (a) ball specimen, (b) disk specimen; contact pressure 2068 MPa, room temperature, rolling speed 3.8 m/s.

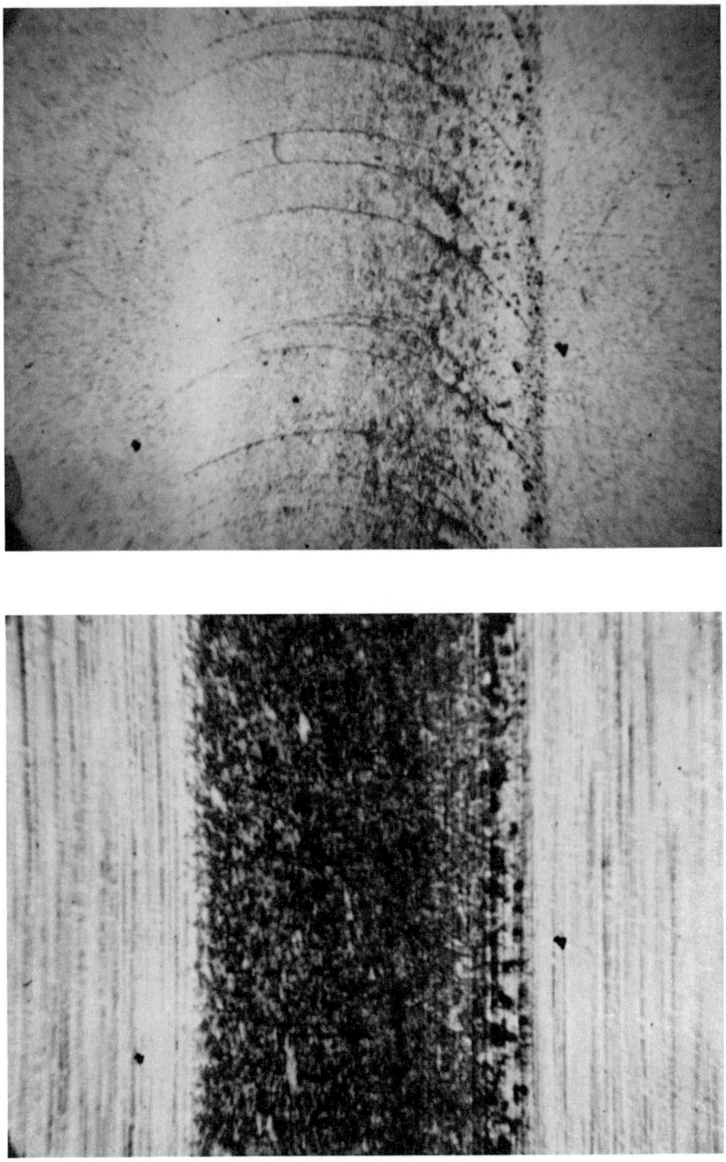

A

B

Fig. 6. Photomicrographs following traction test with (a) silicon nitride ball and (b) 440C stainless steel disk; contact stress 2068 MPa, room temperature, rolling speed 3.8 m/s.

4.16 SOME THERMAL IMPLICATIONS ON THE LIFE OF HIGH SPEED ROLLING ELEMENT BEARINGS

Richard Nicholson

INTRODUCTION

Within Rolls-Royce, the subject of rolling contact fatigue and lubricated wear have been topics of interest for many years. Gear scuffing and roller bearing slip/wear were driving problems during the 1960's, with Smith [1] investigating the parameters contributing to wear in high speed gas turbine roller bearings. With the lessening of our interest in geared transmission systems during the 1970's, the gear work lost its priority, and solutions were found to the roller bearing slip/wear problems. At the same time, the possibility of fracture failure modes in rolling element bearings resulting from higher race hoop stresses became apparent.

Currently, geared fan and propfan systems are being seen as more attractive with advanced propeller aerodynamics becoming available. In addition, system temperatures have risen with consequential reduction in oil film thicknesses, and mixed lubrication failure modes are now also seen in rolling element bearings.

Thus, gear scuffing and mixed lubrication failures in rolling element bearings, roller bearing slip/wear, and combined stress race failure modes are all seen as problems needing attention.

With regard to improved materials, industrial efforts are hindered by lack of a fundamental understanding of the dynamics of the electrohydrodynamic (EHD) contacts. Financial and timescale constraints prohibit the selection and development of a material from an extensive database of material test history, in the way it has been done in the past. There is a need to "design" the next generation of high performance materials to achieve the prescribed duty. Therefore, the duty must be characterised more precisely, including comprehensive mechanical and thermal stress descriptions, and a better knowledge of the fatigue/wear process.

As a part of some more recent investigations into high speed ball bearing performance, a model of the thermal environment was generated. It was based on the power distribution within a bearing being dissipated through a simple heat flow network [2]. This has led to the consideration of transient thermal effects associated with the passage of each rolling element.

THEORETICAL CONSIDERATIONS

As each element passes, there will be a temperature rise associated with the moving heat pulse. Three methods have been tried to estimate the magnitude of the transient temperature rise.

(a) The heat flow network [2] provides a knowledge of the steady state temperature of the raceways and surrounding oil. This allows an estimate of the temperature excursions to be made, by consideration of the mark-space ratio of the heat pulse from the ball.

(b) Winer [3] presents a formula for predicting the peak temperature under rolling and sliding contacts. Using the equation quite considerably outside of the range of the test data gives broad agreement with the above estimate.

(c) Barber [4] presents equations to describe the temperature and stress resulting from a moving linear heat source. The magnitude of the heat source can be obtained by dividing the Hertzian pressure distribution into strips in the rolling direction, and applying a friction coefficient to the power in that strip (Fig. 1), i.e.

$$W = \mu \times Vs \times dx \times \pi \times P' \times a'/2$$

where W = strength of the heat source, Vs = sliding velocity, dx = width of strip in the rolling direction, P' = maximum Hertzian pressure in that strip, a' = width of strip in transverse direction, μ = friction coefficient.

The Barber equations may then be numerically integrated over the contact region (and for some distance before and behind the contact) to produce surface temperatures and stresses (Fig. 2 and Fig. 3).

The magnitude of the temperature rise will be effected by the rolling and sliding velocities, and the Hertzian pressure, as well as material properties. The subsurface temperature distribution has been estimated by a relaxation method applied to a one-dimensional heat flow solution. In high rolling speed applications, this temperature will not penetrate very far into the surface. However, its magnitude can be quite significant. Temperature rises of about 100K (180°F) can be predicted. This will set up an additional thermal stress system, the magnitude of which has not yet been evaluated. However, a correlation has been undertaken comparing the lives from many bearing tests (including combined stress fields) with various lifing criteria. It was noted that the inclusion of a parameter proportional to the peak transient thermal stress (transient temperature rise) reduced the scatter in the ratio of observed life/calculated life by a factor of 3. It therefore seems to be a factor contributing to the life reduction in aero gas turbine ball bearings. It should be noted that all of the bearings considered operate with low Λ values (film thickness/surface roughness).

DISCUSSION

So far, it is not clear why the transient temperature affects the life achieved. The next part of the investigation is to calculate the magnitude of the subsurface thermally induced stresses. The depth of these stresses will only be a small proportion of the Hertzian semi-axis, but there may be regions where the location of critical stresses may move. This is most likely under conditions of high traction and high sliding. This could prove to be a key factor contributing to the fatigue/wear characteristics of such systems.

REFERENCES

1. "Some Aspects of the Performance of High Speed Lightly Loaded Cylindrical Roller Bearings," C. F. Smith, *Proceedings of the Institute of Mechanical Engineers*, *176*, 566-581 (1962).
2. "The Prediction of the Operating Temperatures in High Speed Angular Contact Ball Bearings," R. Nicholson, to appear in *Proceedings of 13th Leeds-Lyon Symposium on Tribology*, Leeds (1986).
3. "Surface Temperature Measurements in Rolling and Sliding EHD Contacts," H. S. Nagaraj, D. M. Sanborn and W. O. Winer, *American Society of Lubrication Engineers Transactions*, *22*, 227-285 (1979).).
4. "Thermoelastic Displacements and Stresses Due to a Heat Source Moving Over the Surface of a Half Plane," J. R. Barber, *Journal of Applied Mechanics, Transactions of the American Society of Mechanical Engineers*, *51*, 636-640 (1984).

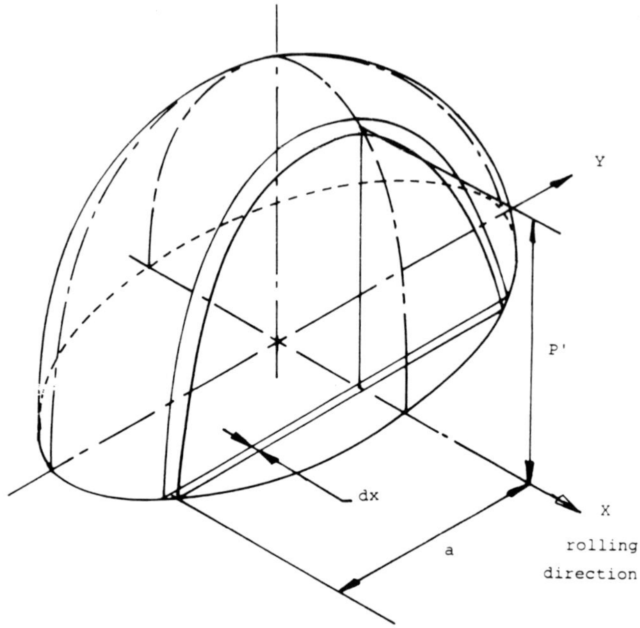

Fig. 1. Hertzian Pressure Profile.

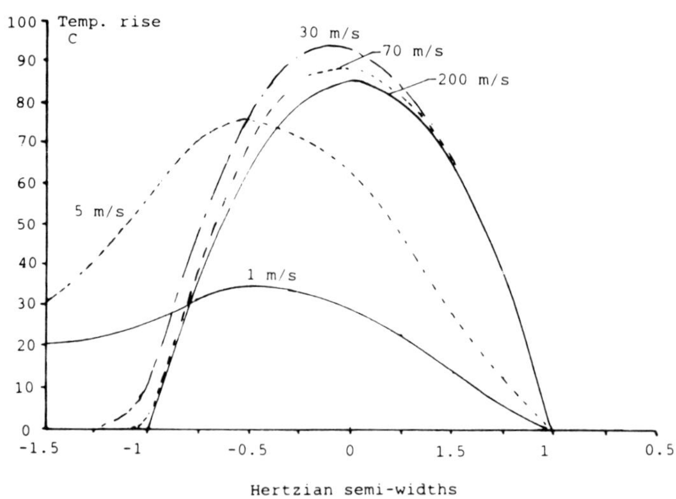

modulus	2.14E11 Pa	expans coefft.	1.0 E-5
density	8680 kg/m^3	specific heat	508 J/kg K
max. Hertz press	1.38E9 Pa	friction coefft.	0.025
Hertz semi-axes	4.26E-3 m	rolling vel.	50 m/s
	2.96E-4 m		

Fig. 2. Surface Temperature Rise for Various Sliding Velocities.

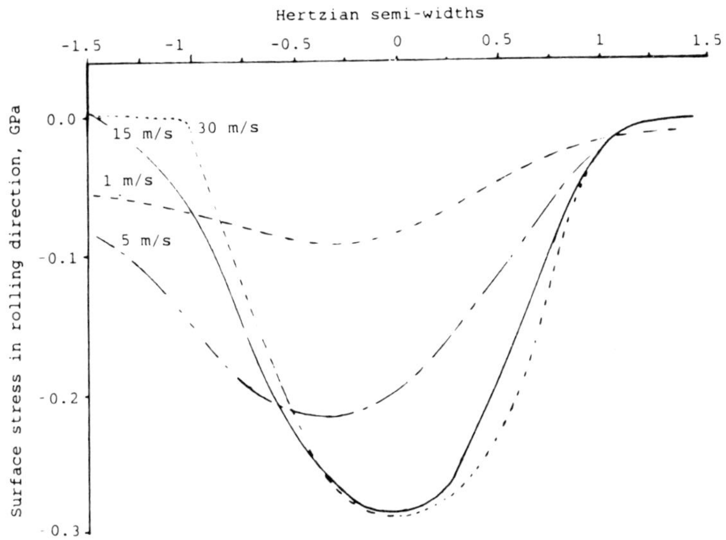

modulus 2.14E11 Pa
density 8680 kg/m^3
max. Hertz press 1.38E9 Pa
Hertz semi-axes 4.26E-3 m
2.96E-4 m
expans coefft. 1.0 E-5
specific heat 508 J/kg K
friction coefft. 0.025
rolling vel. 50 m/s

Fig. 3. Surface Stress for Various Sliding Velocities.

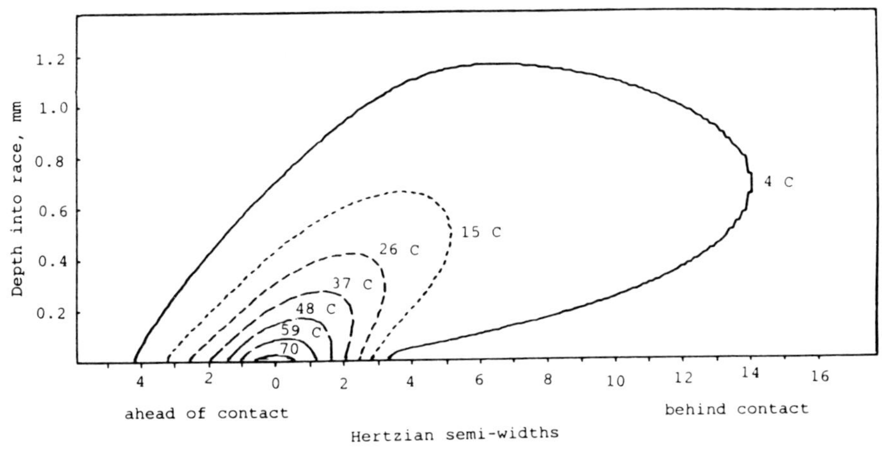

modulus 2.14E11 Pa
density 8680 kg/m^3
max. Hertz press 1.38E9 Pa
Hertz semi-axes 4.26E-3 m
2.96E-4 m
expans coefft. 1.0 E-5
specific heat 508 J/kg K
friction coefft. 0.025
rolling vel. 50 m/s
sliding vel. 5 m/s

Fig. 4. Subsurface Temperature Profile.

4.17 PROFILOMETRIC ROUGHNESS AND CONTACT FATIGUE

Tibor E. Tallian

INTRODUCTION

Contact fatigue life prediction models have existed for half a century [1]. Recent refinements take into account the microstresses superimposed on the Hertz contact stress field as a result of roughness in the two contacting surfaces [2,3,4]. These latter efforts are characterized by the use of some simple asperity geometry (rounded-top prisms in [2] and Greenwood-Williamson models using Hertzian second-order) asperity shapes in [4]), to calculate microstress fields under asperities. Given these stress fields, the models predict fatigue life (in cycles) by integration, over the stressed volume, of a relationship of the form:

$$- \ell n(dS) = Ct^a z^b N^c dV \qquad (1)$$

where C is a constant, t a stress, z a linear dimension, dV a volume element and dS a local survival probability. a, b and c are (positive or negative) numeric exponents.

With the progressive refinement of surface microgeometry description, of stress analysis for contacts between arbitrary surfaces, and of contact fatigue life modeling, the use of prescribed-shape asperities for microstress computation is becoming one of two weak links in the modeling effort. The other weak link may be identified as the adherence to the Palmgren-Miner linear damage accumulation hypothesis represented by Eq. (1), which no longer appears to best represent the phases of fatigue damage accumulation [5]. This writer is insufficiently grounded in fracture mechanics to address the *second* shortcoming. This note discusses the *first*, i.e. the asperity geometry model.

PROFILOMETRIC SURFACE GEOMETRY DESCRIPTION

A desirable feature of a model describing surface roughness is the absence of a priori shape assumptions for its elements. A well known model with this feature is the profilometric one, based on Longuet-Higgins' description of a two-dimensional random process [6], which has been adapted to rough surfaces by Nayak and is extensively described in several papers by McCool [7]. This model uses the spectral moments of the two-dimensional height distribution of a surface as parameters in the description of its statistics. In its general form, this model has few stringent assumptions, and none are geometrical, such as asperity tip shapes.

Longuet-Higgins profilometric models have seen extensive use in modeling partial elastohydrodynamic lubrication [8], but little if any, in modeling microstresses between rough surfaces. The analytical difficulty of a stress analysis without Hertzian assumptions for the shape of asperities, is well recognized.

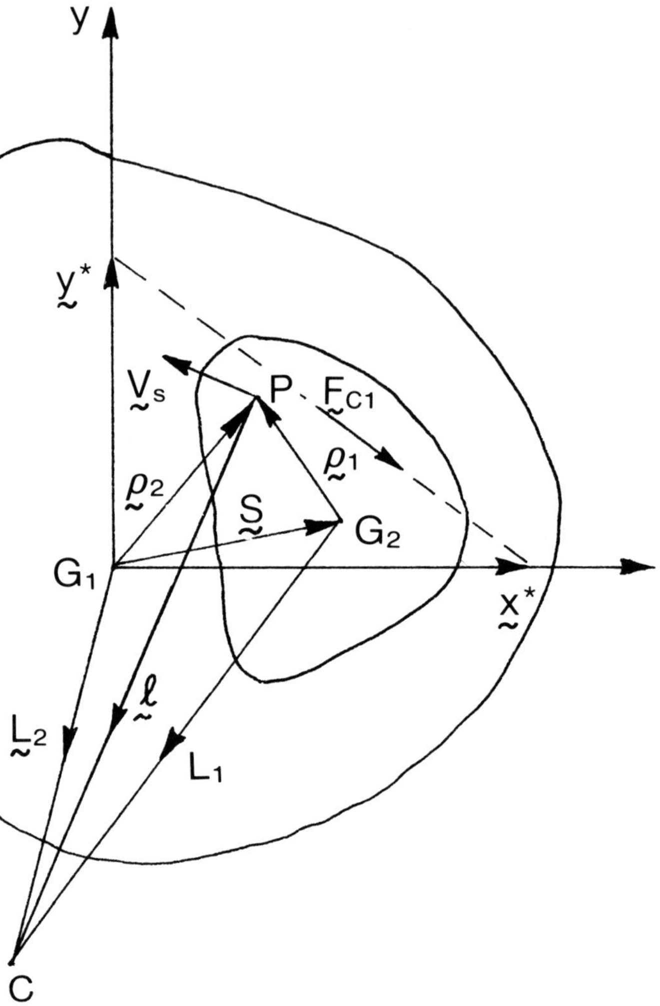

Fig. 1. Two Bodies in Friction Contact.

4.19 MICROSCOPIC MECHANISMS OF TRIBOLOGICAL AND WEAR PROCESSES: MOLECULAR DYNAMICS SIMULATIONS

Martin W. Ribarsky and Uzi Landman

Although tribological phenomena are an everyday experience and have been observed and studied for a very long time [1], a detailed microscopic theory of tribological phenomena is lacking. Nevertheless a large body of empirical data has been collected and some phenomenological models have been developed [2]. The major reasons, from a theorist's perspective, for this state of affairs may be attributed to: (i) the complexity of the phenomena, which involve processes occurring in materials under stress, the generation and propagation of defects, large structural deformations and properties of systems beyond the elastic regime, (ii) the nature of available experimental data, which until fairly recently was not obtained under controlled conditions on compositionally and structurally well characterized samples. Recent advances in computer simulation methods [3] and the advent and application of surface science experimental probes provide the impetus for the development of fundamental models, on the atomic scale, of tribological and wear phenomena. In this note we outline our recent investigations in this area using molecular-dynamics (MD) simulations, and demonstrate the potential of such investigations towards the development of a microscopic understanding of the physical processes which underly these phenomena.

METHOD

Computer simulations, where the evolution of a physical system is simulated with refined temporal and spatial resolution, via direct numerical solution of the equations of motion of a system of interacting particles, are in a sense computer experiments which open new avenues in investigations of the microscopic origins of material phenomena [3]. In its simplest form [4], the interaction between particles i and j is described by a pair potential $\phi(r_{ij})$ where $r_{ij} = |\vec{r}_i - \vec{r}_j|$, and the classical equations of motion for a system of N particles, of masses M_i, are given by

$$M_i \frac{d^2\vec{r}_i}{dt^2} = - \sum_{j=i}^{N} \frac{\partial \phi(r_{ij})}{\partial \vec{r}_i} \quad . \tag{1}$$

Practical considerations restrict the number of dynamical particles contained in the calculational cell to be of the order of 10^3-10^4, and the simulation of the extended system is achieved via the employment of periodic boundary conditions. For exploration of the elastic behavior of solids and phenomena which may involve external stresses, volume changes, shape and structural deformations, it is important that the variables which determine the volume and shape of the calculational cell will be allowed to respond dynamically. In our calculations we have used a newly developed method [5] which allows for the above dynamical considerations.

MODEL CALCULATIONS

To illustrate the method we have constructed a model system consisting of 1200 (or 1260) particles in the calculational cell (which is periodically repeated), interacting via 6-12 Lennard-Jones potentials,

$$\phi_{\alpha\beta}\ (r) = 4\epsilon_{\alpha\beta}\ [(\sigma_{\alpha\beta}/r)^{12} - (\sigma_{\alpha\beta}/r)^{6}] \qquad (2)$$

with $\sigma_{\alpha\beta} = (\sigma_{\alpha\alpha}+\sigma_{\beta\beta})/2$ and $\epsilon_{\alpha\beta} = (\epsilon_{\alpha\alpha}\ \epsilon_{\beta\beta})^{1/2}$. The bottom half of the system (α) was characterized by an interaction strength parameter $\epsilon_{\alpha\alpha} = 2\epsilon_{\beta\beta}$, where $\epsilon_{\beta\beta}$ determines the interaction strength in the other part of the system (β). In this arrangement the soft material (β) may be viewed as a solid lubricant, whose melting point is half that of the surrounding material. In order to isolate the effect of interaction strength from other material parameters we have taken $\sigma_{\alpha\alpha} = \sigma_{\beta\beta}$. In the following we use reduced units [6] where energy (and temperature) are expressed in units of $\epsilon_{\alpha\alpha}$, length in units of $\sigma_{\alpha\alpha}$ and the time unit (t.u.) is $(m_\alpha/\epsilon_{\alpha\alpha})^{1/2}\sigma_{\alpha\alpha}$, (the integration time-step Δt was 0.0075t.u.). Two series of studies were conducted: (1) The interface between the hard (α) and soft (β) components of the system was in the (001) plane, $N_{layer} = 20$, $N = 1200$; layers 1-10 of material α and 11-20 of material β. (2) The interface was in the (111) plane, $N_{layer} = 18$, $N = 1260$; interface between layers 9 and 10. At the start of the calculations the systems were equilibrated at a reduced temperature $T \sim 0.11$ (about $T^{(\alpha)}_{melting}/6$). Subsequently, a load [6] of 0.5 was imposed along the direction normal to the interface. Following equilibration under load, the system was subjected to shear stress (in the <110> direction for the (001) interface, and in the $<1,\bar{1},0>$ direction for the (111) interface) which was increased gradually until the system failed. Special care was taken to determine the yield stress, σ_c. In addition, simulations at constant external shear-stress were conducted. In this report we restrict ourselves to thermally isolated (adiabatic) systems.

From the temporal evolution of the particle trajectories, generated via the integration of the equations of motion, the internal stress tensor in the system can be calculated [7]. The time evolution of the internal stress in layers, σ_{xz} (i.e., along the <110> direction), for a fixed external stress of 1.9375 ($\sigma_c = 1.925 \pm 0.012$), just before and during yield for the (001) system, is shown in Fig. 1a. We observe that variations in the internal stresses start in the soft material, in a region removed from the interface (originally located between layers 10 and 11), i.e., layers 13-15, while layers 11 and 12 which are adjacent to the interface are pinned, at the onset of yield, by their interaction with the hard substrate.

This sequence of events is further demonstrated by the particle trajectories shown in Fig. 2. Here, the initial position of an atom is at the center of the circle, from where the indicated trajectory develops. These trajectories correspond to the time interval 232-235 t.u. (see Fig. 1a). As seen, while layers 11 and 12 move uniformly with layer 10 of the substrate, deeper layers in the soft material do not. Note that the movement in either direction tends to involve entire rows rather than having motion in several directions for some rows. In fact, layers 16, 17, 18, 19, and 20 viewed together show that, in successive planes, one new row at a time joins in the leftward movement until all rows are moving to the left by layer 20. A geometrical analysis of the slip in layers 16 through 20 reveals that two intersecting slip planes are involved: $\{1\bar{1}1\}$ and $\{\bar{1}11\}$. Thus, even though the system was sheared along the {001} plane in the (110) direction, the slip occurred in {111}-type planes. This is just the expected experimental result since all methods consistently indicate that {111}-type planes in (110) directions are the major operative slip systems in FCC structures. The deformation of the system is accompanied by an increase in temperature (see Fig. 1b), which "autocatalyzes" plastic deformation and eventual total yield. The coefficient of friction which is obtained from these calculations σ_c/load = 4, which is of the order of typical values for the static coefficient for clean metal surface in contact in vacuum.

To demonstrate the dependence of the mechanisms of deformation and transformation on the crystallography of the interface, we have performed similar simulations for a system in which the interface was in the (111) plane. We find that under adiabatic conditions this system yields at $\sigma_c = 0.859$. When the system is sheared in the $\langle 1\bar{1}0 \rangle$ direction, under adiabatic conditions, below the yield stress, it slips at $\sigma_s = 0.797 \pm 0.016$ generating stacking faults in the soft (lubricant) material. Allowing the system to relax under σ_s and then increasing the imposed stress, the system yields at $\sigma_c = 0.859 \pm 0.016$. Similar calculations for a system under isothermal conditions exhibit a slip stage occurring at $\sigma_s = 0.85 \pm 0.025$ and eventual yield at $\sigma_c = 0.97 \pm 0.03$. The system in contact with a heat reservoir (isothermal) withstands higher shear stresses as compared to the adiabatic case, since in the latter case the heat which is generated in the course of the deformation of the system is utilized to overcome potential barriers for sliding, while in the former case this heat is dissipated to the reservoir and higher shear stresses are necessary to bring about a similar effect. In addition, we note the difference between the critical stresses σ_c, for the (111) and (001) interface orientations. The degree of detail afforded by the MD simulations is further demonstrated by the selected particle configurations shown in Fig. 4, for the adiabatic (111) interface, exhibiting the formation of a stacking fault in the soft material. For this system the original interface is between layers 9 and 10 and the stacking fault is generated at layers 12 and 13 (see Fig. 3, in which a slice of the system viewed along the $\langle 10\bar{1} \rangle$ direction, before and after the generation of the stacking fault, is shown). The sequence of configurations shown in

Fig. 4 (separated by 170 Δt) show layers 12 and 13 starting at registry (4a) and moving out of registry and into new equilibrium positions Fig. 4c), generating a stacking fault.

While in the above we focused on systems characterized by interfaces with differing interaction strengths, the effect of other material parameters such as lattice parameter mismatch, covalent (directional) bonding and long-range interactions (as found in ionic materials) as well as effects due to ambient conditions such as shear rates, and normal load values, are under investigation. Finally, while we found evidence for the generation of dislocations in our simulations, the full development of these extended defects would require simulation for larger systems.

ACKNOWLEDGEMENTS

This work is supported by NSF Grant No. MSM-8611257 and by a contract from the Hughes Aircraft Company. Partial support was provided by U.S. DOE Grant No. F505-86ER45234.

REFERENCES

1. "Sur les frottements des machines," G. Amontons, *Histoire de l'Academie Royale, 206*, 128-138 (1699).
2. *Surface Effects in Adhesion, Friction, Wear and Lubrication*, D. H. Buckley, Elsevier, New York (1981), and references therein.
3. "Microscopic Phenomena of Macroscopic Consequences: Interfaces, Glasses, and Small Aggregates," U. Landman, R. N. Barnett, C. L. Cleveland, W. D. Luedtke, M. W. Ribarsky, D. Sharf and J. Jortner, in *Computer-Based Microscopic Description of the Structure and Properties of Materials*, *MRS Symposia Proceedings, 63*, 273-284 (1986).
4. For a formulation and MD simulations beyond the pair-potential approximations see "New Molecular-Dynamics Method for Metallic Systems," R. N. Barnett, C. L. Cleveland and U. Landman, *Phys. Rev. Lett. 54*, 1679-1682 (1985) and "Faceting at the Silicon (100) Crystal-Melt Interface: Theory and Experiment," U. Landman, W. D. Luedtke, R. N. Barnett, C. L. Cleveland and M. W. Ribarsky, *Phys. Rev. Lett. 56*, 155-158 (1986).
5. "Polymorphic Transitions in Single Crystals: A New Molecular Dynamics Method," M. Parrinello and A. Rahman, *Journal of Applied Physics, 52*, 7182-7190 (1981).
6. If we choose interaction parameters which correspond to the cohesive energy and lattice constant of Nickel ($\epsilon = 3.64\times10^{-13}$erg, $\sigma = 2.49$Å and $M = 9.75\times10^{-23}$g) a reduced temperature $T = 0.11$ corresponds to 300°K, the melting temperature $T_m = 0.7$ to 2000°K, the reduced unit of stress or load to 2.4×10^{7}g/cm^2 (or 2.4×10^{10} dynes/cm^2) and the time-unit corresponds to 4.1×10^{-13} sec. Time expressed in terms of the above time unit is denoted by t^*.
7. "Local Structural Fluctuations in Amorphous and Liquid Metals: A Simple Theory of the Glass Transition," T. Egami and D. Srolovitz, *Journal of Physics of Fluids, 12*, 2141-2163 (1982).

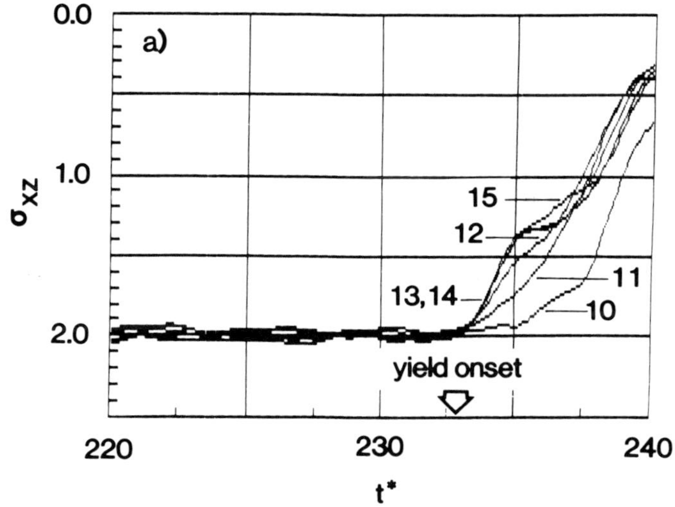

Fig. 1a. Internal shear stress per layer in the same direction and along the same plane as the external stress, σ_{ext} = 1.9375, versus reduced time. Note the initiation of response to the applied stresses in layers 13-15.

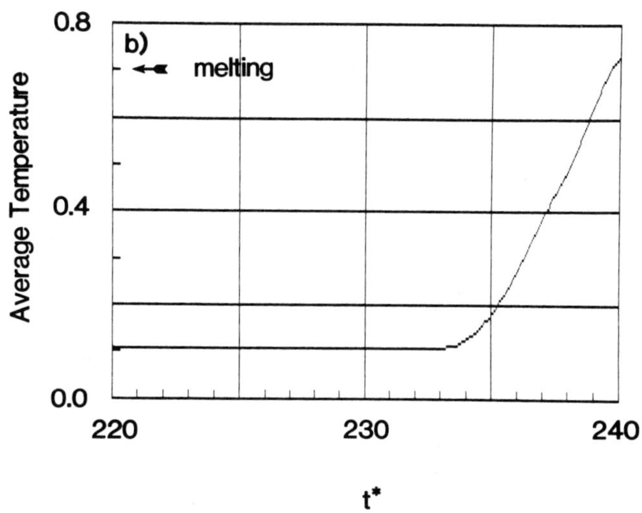

Fig. 1b. Temperature for the total system versus reduced time.

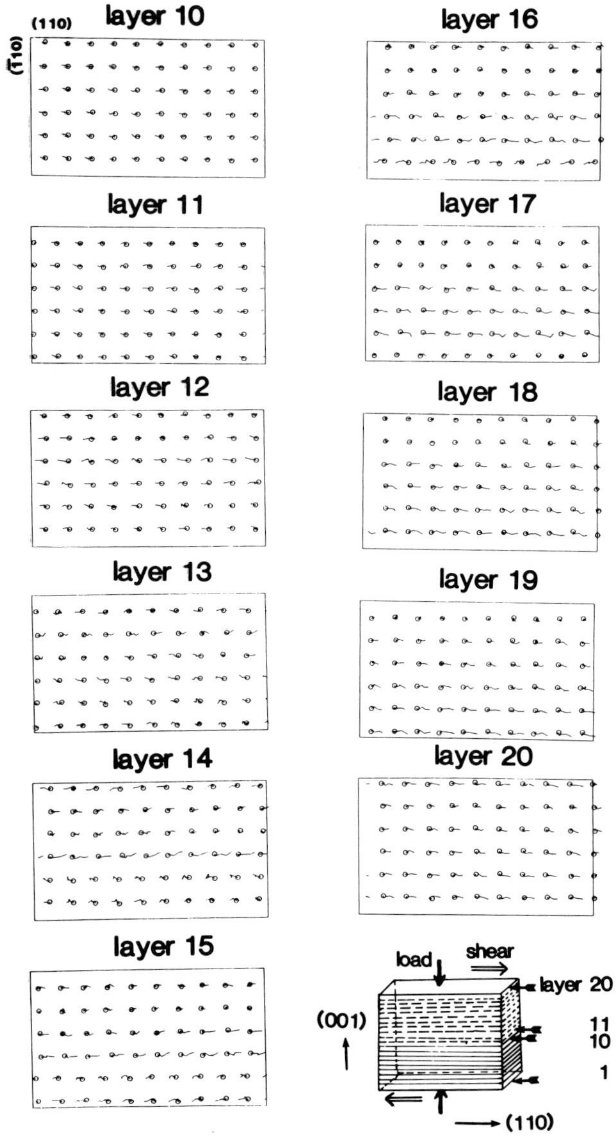

Fig. 2. Atomic trajectories for successive {001} layers. The hard-soft interface is between layers 10 and 11. The trajectories shown are for σ_{ext} = 1.9375 at t* = 232-235 t.u. A schematic of the calculational cell is shown at the bottom right hand side.

However, given the powerful finite element methods of stress analysis now at hand, it would seem timely to re-examine the feasibility of building a profilometrically based microstress model for use in contact fatigue life prediction.

Such a model should comprise the following elements:

1. Describe the two contacting surfaces profilometrically.
2. Allow for the interposition of an EHD film.
3. Calculate microstresses caused by roughness, using no auxiliary assumptions as to asperity geometry.
4. Superimpose Hertzian macrostresses on these microstresses.
5. Select a physically supported critical stress criterion.
6. Invoke a fatigue damage accumulation model (as in Eq. (1) until a more general model becomes available), for each stressed volume element and calculate failure probability for the total stressed volume by integration.

Numerous refinements, such as inclusion of traction effects on the stress field, residual stresses, non-zero fatigue limit stress, and consideration of non-Hertzian macrocontact geometries, can be added.

PAPER BY SEABRA AND BERTHE

In a recent paper [9] these authors present a stress analysis of rough contact surfaces free from assumptions about asperity shape. Their results are, as of this paper, restricted to one wavelength at a time and cover only surface pressures, not sub-surface stresses. However, their solution algorithm does not depend on a single-wavelength restriction. Numeric calculation of the subsurface stress field from a known arbitrary surface pressure distribution is an established procedure [10]. Thus, all stress analysis steps in the profilometric approach suggested for the proposed model structure have been individually traversed and can be made part of a unified contact fatigue prediction method.

GOAL AND JUSTIFICATION

The eventual goal of the proposed model is to produce a truly unified description of contact fatigue damage accumulation.

Uncoordinated development of the many sub-elements of the contact fatigue model has, by now, led to contact fatigue life prediction methods, which are both difficult to penetrate and to validate. As more and more, individually meritorious, refinements are added, the situation becomes increasingly obscure. Of particular concern are possible contradictory effects from restrictive assumptions, when they differ in the different submodels. Thus, the EHD submodel does not use Greenwood-Williamson asperity shapes, whereas the stress field model does. Might not Greenwood-Williamson asperities behave startlingly as EHD elements? Another contradiction is the discretization of asperities for stress modeling, and the need for volume integration in calculating damage accumulation. How should the volume elements "between" asperities be treated?

Problems of possible internal inconsistency also arise in the damage accumulaation model, when fatigue limit stresses are postulated, or crack-propagation terms are added, while maintaining Palmgren-Miner linear damage accumulation. It can be expected, that the fracture mechanics community will soon present us with more sophisticated and physically more meaningful damage accumulation models than those now in use [11]. With every refinement in the damage accumulation model, the need for a maximally assumption-free microstress model increases.

REFERENCES

1. "Dynamic Capacity of Rolling Bearings," G. Lundberg and A. Palmgren, *Acta Polytechnicha*, *1*, No. 3, 1-50 (1947).
2. "Prediction of Traction and Microgeometry Effects on Rolling Contact Fatigue," T. E. Tallian, Y. P. Chiu and E. van Amerongen, *Journal of Lubrication Technology, Transactions of the American Society of Mechanical Engineers*, *100*, 156-166 (1978).
3. "Unified Rolling Contact Life Model with Fatigue Limit," T. E. Tallian, *WEAR*, *107*, 13-36 (1986).
4. "Elastic Stresses Below Asperities in Lubricated Contacts," E. Ioannides and J. C. Kuypers, *Journal of Tribology, Transactions of the American Society of Mechanical Engineers*, *108*, 394-402 (1986).
5. "Response to Rolling Contact Loading," A. P. Voskamp, *Journal of Tribology, Transactions of the American Society of Lubrication Engineers*, *107*, 359-378 (1985).
6. "The Statistical Analysis of a Random Moving Surface," M. S. Longuet-Higgins, *Philosophical Transactions of the Royal Society, London, Series A*, 249 (1957).
7. "Relating Profile Instrument Measurements to the Functional Performance of Rough Surfaces," J. I. McCool, *ASME Preprint 86-Trib-9* (1986).
8. "The Theory of Partial Elastohydrodynamic Lubrication," T. E. Tallian, *WEAR*, *21*, 49-101 (1972).
9. "Influence of Surface Waviness and Roughness on the Normal Pressure Distribution in the Hertzian Contact," J. Seabra and D. Berthe, *ASME Preprint 86-Trib-14* (1986).
10. *Contact Mechanics*, K. L. Johnson, Cambridge University Press (1985).
11. "Pitting Model for Rolling Contact Fatigue," L. M. Keer and M. D. Bryant, *Journal of Lubrication Technology, Transactions of the American Society of Mechanical Engineers*, *198*, 198-205 (1983).

4.18 TWO-DIMENSIONAL DYNAMICS OF COULOMB FRICTION

Iradj G. Tadjbakhsh

INTRODUCTION

The foundation of our ideas about dry friction were put forth by Amontons (1699) and Coulomb (1785), who postulated that the primary phenomenon in dry friction is the interaction of rough surfaces. Recently Villagio [1] argued that if the contacting bodies deform elastically, then Coulomb's or Amontons' law should not be accepted as a postulate but as a consequence of both the topology and the elastic deformation of the contact surfaces. These basis considerations are the subjects of several current investigations with potential near-future applications.

The laws of friction as enunciated by Coulomb state that the frictional forces oppose relative velocity of the contacting bodies and that the magnitude of these forces are proportional to the interfacial normal stresses. Ample experimental evidence exists to support these statements for relatively high velocities.

ANALYSIS

Let us consider the general plane motion of two bodies with masses m_1 and m_2 which have an area of contact A in the plane of motion. Also let us consider an inertial frame xyz with the z axis normal to the plane of motion. Let G_i and $\underset{\sim}{\omega}_i = \omega_i(t)\underset{\sim}{k}$, i=1,2, denote the projections of the mass centers on the plane of motion and the angular velocities of the two bodies. As shown in Fig. 1, the position of the instantaneous center of relative motion of m_1 with respect to m_2 is called C with coordinates $x_c(t)$ and $y_c(t)$.

The resultant frictional force and couple on m_1 are given by

$$\underset{\sim}{F}_{fi}(t) = \iint_A \underset{\sim}{\tau}_i \, dA \tag{1}$$

$$\underset{\sim}{M}_{fi}(t) = \iint_A (\underset{\sim}{L}_i - \underset{\sim}{\ell}) \times \underset{\sim}{\tau}_i \, dA \tag{2}$$

where $\underset{\sim}{\tau}_i$ is the shear stress of a generic point P of A that acts on the body m_i and the vectors $\underset{\sim}{\ell}(t)$ and $\underset{\sim}{L}_i(t)$ are as defined in Fig. 1.

The shear stresses $\underset{\sim}{\tau}_i$ depend upon the normal pressure p(x,y,t), the relative local velocity $V_s(x,y,t)$ as well as the physical properties of the contacting surfaces. For prescribed surfaces $\underset{\sim}{\tau}_i$ can be represented by

$$\underset{\sim}{\tau}_1 = -\underset{\sim}{\tau}_2 = \operatorname{sgn}(\omega_s) f(p,V_s)\underset{\sim}{k} \times (\underset{\sim}{\ell}/\ell) \tag{3}$$

where the function $f(p,V_s)$ contains the essential constitutive information on behavior of dry friction. A possible variation of f with V_s and p and other relevant properties are subjects of several on-going investigations.

Let us consider a point of body m_1 and a point of body m_2 that at an instant t are in contact with each other at a point P of A. The velocities $V_1(x,y,t)$ and $V_2(x,y,t)$ of these points are given by

$$\underset{\sim}{V}_i = \underset{\sim}{v}_i + \underset{\sim}{\omega}_i \times \underset{\sim}{\rho}_i \tag{4}$$

where $\underset{\sim}{v}_i(t)$ are the velocities of the mass centers and $\underset{\sim}{\rho}_i(x,y,t)$ are as shown in Fig. 1. Then with reference to C

$$\underset{\sim}{V}_s = \underset{\sim}{\ell} \times \underset{\sim}{\omega}_s \tag{5}$$

These equations also imply [2]

$$\underset{\sim}{\ell} = \frac{1}{\omega_s} \underset{\sim}{k} \times \underset{\sim}{V}_s = \frac{1}{\omega_1 - \omega_2} (\underset{\sim}{k} \times \underset{\sim}{v}_s - \omega_1 \underset{\sim}{\rho}_1 + \omega_2 \underset{\sim}{\rho}_2) \tag{6}$$

$$\underset{\sim}{L}_1 = \frac{1}{\omega_1 - \omega_2} (\underset{\sim}{k} \times \underset{\sim}{v}_s + \omega_2 \underset{\sim}{S}) \tag{7}$$

$$\underset{\sim}{L}_2 = \frac{1}{\omega_1 - \omega_2} (\underset{\sim}{k} \times \underset{\sim}{v}_s + \omega_1 \underset{\sim}{S}) \tag{8}$$

where

$$\underset{\sim}{S} = \underset{\sim}{L}_2 - \underset{\sim}{L}_1 \tag{9}$$

When v_i and ω_i are known, the foregoing enable one to determine $\underset{\sim}{F}_{fi}$ and $\underset{\sim}{M}_{fi}$ from (1) and (2). When v_i and ω_i are unknown then (1) - (2) imply that $\underset{\sim}{F}_{fi}(t) = \underset{\sim}{F}_{fi}(v_i(t),\omega_i(t))$ and $\underset{\sim}{M}_{fi}(v_c(t),\omega_i(t))$ and Newton's laws have to be invoked

$$\underset{\sim}{m}_i \ddot{v}_i = \underset{\sim}{F}_i + \underset{\sim}{F}_{fi} + (-1)^i \underset{\sim}{F}_{12}\ , \qquad i=1,2 \tag{10}$$

$$I_i \dot{\underset{\sim}{\omega}}_i = \underset{\sim}{M}_i + \underset{\sim}{M}_{fi} + (-1)^i \underset{\sim}{M}_{12}\ , \qquad i=1,2 \tag{11}$$

Here $\underset{\sim}{F}_i$ and $\underset{\sim}{M}_i$ are the externally applied forces and couples and $\underset{\sim}{F}_{12}$ and $\underset{\sim}{M}_{12}$ are inter-body forces and couples that may arise as a result of the action of connecting elements such as massless springs or cords.

When the two bodies move as a unit (sticking) but are at incipient slipping frictional forces and couples cannot be determined from (1) - (2). Denoting these by $\underset{\sim}{F}'_{fi}$ and $\underset{\sim}{M}'_{fi}$ they can be determined from (10) - (11) when one supplements them with the additional information

$$\underset{\sim}{\omega}_1 = \underset{\sim}{\omega}_2 = \underset{\sim}{\omega} \tag{12}$$

$$\underset{\sim}{v}_1 = \underset{\sim}{v}_2 = \underset{\sim}{\omega} \times \underset{\sim}{S} \tag{13}$$

$$\dot{\underset{\sim}{v}}_1 = \dot{\underset{\sim}{v}}_2 + \dot{\underset{\sim}{\omega}} \times \underset{\sim}{S} - \omega^2 \underset{\sim}{S} \tag{14}$$

It was shown in [2] that

$$\tilde{F}'_{fi} = \frac{1}{2+\beta S^2}(\tilde{f}+\beta \tilde{M}\times\tilde{S}) + \frac{\beta}{2(2+\beta S^2)}(\tilde{f}\cdot\tilde{S})\tilde{S} - \frac{m_1}{2}\omega^2\tilde{S} \tag{15}$$

$$\tilde{M}'_{fi} = \tilde{M}_{12} - \left\{ \frac{I_2}{I} + \frac{\beta S^2 I_1}{(1+\beta S^2)I} \right\} \tilde{M}_1 + \frac{I_1(\tilde{M}_2 - \tilde{S}\times\tilde{f})}{I(1+\beta S^2)} \tag{16}$$

where

$$\beta = \frac{m_1}{I_1+I_2}, \quad \tilde{f} = 2\tilde{F}_{12} + \tilde{F}_2 - \tilde{F}_1, \quad I = I_1 + I_2 \tag{17}$$

Slipping takes place when conditions are such that

$$|F'_{f1}| > |F_{f1}| \tag{18}$$

and the converse relationship implies transition to non-sliding motion.

REFERENCES

1. "An Elastic Theory of Coulomb Friction," P. Villagio, *Archive for Rational Mechanics and Analysis*, *70*, 135-143 (1979).
2. "Analysis of General Plane Motion with Coulomb Friction," C. Younis, I. Tadjbakhsh and E. A. Saibel, *WEAR*, *91*, 319-331 (1983).

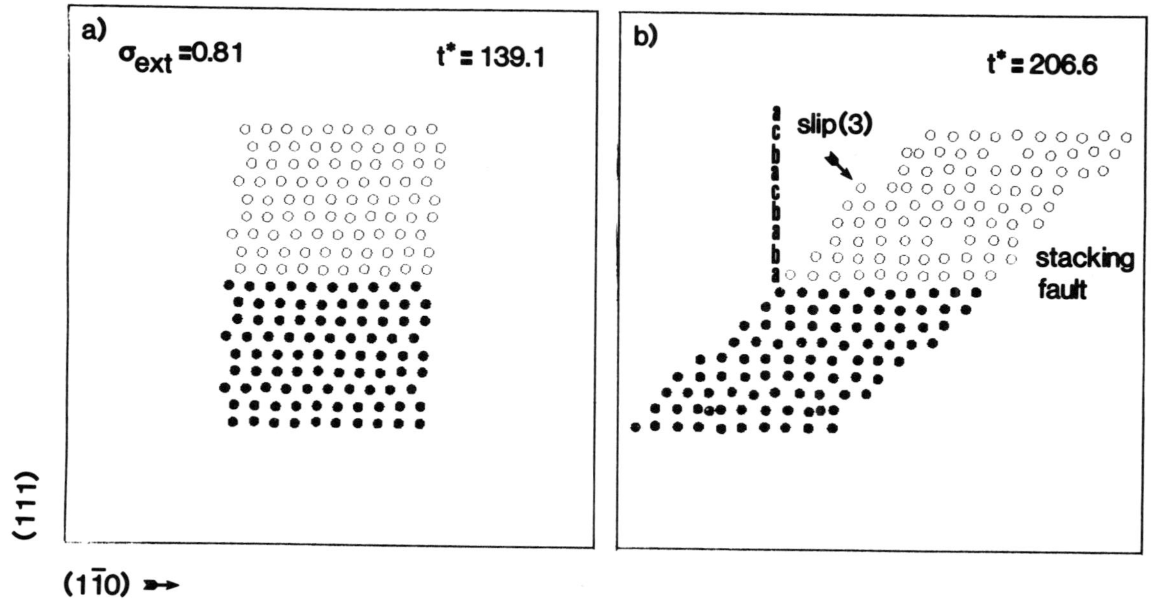

Fig. 3. Atomic trajectories for a central slice of the {111} system at σ_{ext} = 0.81. a) Before slip and b) after slip. A stacking fault occurs in layers 10-13, and layer 15 slips 3 rows to the left.

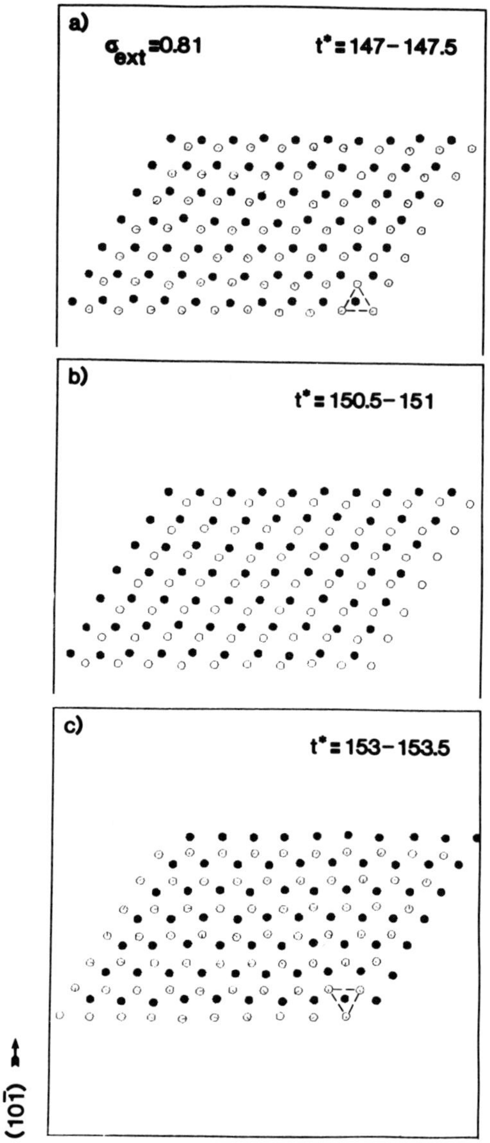

Fig. 4. Layers 12 (shaded circles) and 13 from the {111} system during the formation of the stacking fault. a) Before the stacking fault forms, the registry between layers is indicated by the dotted lines. b) Midway through stacking fault formation. c) The layers have moved to a new registry, indicated by the dotted lines.

4.20 ON FRACTAL DIMENSION OF ENGINEERING SURFACES

Frederick F. Ling and Salvadore J. Calabrese

INTRODUCTION

One of the most important properties of machine elements is their surface roughness. As such, surface roughness profiles, and topography in general, have been measured. Various instruments have been developed for such measurements.

As these instruments become increasingly sophisticated, modeling of surface roughness has not kept pace. Modeling of surface profile per se has not occupied much attention historically. On the other hand, modeling has been carried out in conjunction with assigned engineering functions of surfaces. For example, in micro-elastohydrodynamic lubrication theory the "bump" of a surface asperity may be modeled as a spherical cap so that the elastic deformation of this bump can be related to the assumed geometry via the Hertz theory. Elsewhere in this volume, Ling and Peterson have given a simple model of surface roughness of the die in the modeling of metalworking friction under extreme pressure.

The two examples given above, and many more, are deterministic. That is, a unit event is modeled in the quantitative sense. The number of units involved in a given overall model depends on the accuracy sought for the problem at hand. Seldom is a deterministic model carried out very far, i.e. the bump and its neighbors, then the neighbors of neighbors, etc. The reason is that the associated calculation would soon become so horrendous as to render the job of implementation impractical. To capture the influence of a large number of asperities, surfaces have been modeled probablistically, e.g. in connection with area of contact [1], load-compliance [2], friction [3], and sliding interface temperature transients [4].

As noted before, the modeling of surface roughness per se has not received much attention. Exceptions would include cases where surface roughness height distributions have been modeled as Gaussian and where radii of curvature of peaks of roughness have been modeled as Gaussian. This short paper will address the issues involved in modeling engineering surfaces as they are formed or processed. Needless to say, there are important problems which require the quantitative knowledge of engineering surfaces, and their solutions would be so much more tractable if the surfaces are modeled adequately, i.e. independently of the problems whose solutions require the models. For simplicity, within the spirit of this workshop volume, permit us to dwell only in two dimensions, i.e. the roughness-height as a function of position along a fixed direction on a surface. Much of the ensuing discussion will extend to three dimensions, i.e. the whole surface topography.

EUCLIDEAN DIMENSION AND FRACTAL DIMENSION

In the Introduction, all references to dimensions were based on Euclidean geometry; thus two-dimensional and three-dimensional. In fact, the models were all grounded in Euclidean geometry, e.g. a unit surface asperity modeled as a wedge [1], a cone [2]. During the last two decades, Mandelbrot [5,6] introduced the notion of fractal geometry. He identified a family of shapes which he called fractals. The most useful fractals, he asserts, involve chance and both their regularities and their irregularities are statistical. The concept of fractal dimension plays a central role in describing nature.

Let us examine a section of surface profile and ask the question: how long is the contoured surface along a given direction, i.e. within the two-dimensional Euclidean world? We note, of course, when the surface is mirror flat the profile degenerates into one-dimensional. The answer is "it depends." In other words, given the actual surfaces, the actual length of the surface between two fixed points on a straight line depends on the measuring "stick." When measured by a 1 μm "stick," the length may be L_1, when measured by a 1 nm "stick," the length may be L_2. As the length of the "stick" becomes diminishingly small, the length measured may be infinitely long. The aforementioned may well be true but it is not very useful to say that the surface contour between two markers, say two centimeters apart, is infinitely long. What Mandelbrot proposed is to cut off the two extremes. For example, for the referenced length, one would not use a "stick" which is larger than 1 cm on the one hand, nor would one use a "stick" which is less than 1 Å long on the other hand. Within these limits fractals are invariant under transformation of scale; this fact is called scaling.

Mandelbrot described the Koch curve. As shown in Fig. 1a, this curve has one of the simplest, basic constructions. Starting with a straight segment $\overline{AB}$, which is normalized by a length L, divide it into three equal parts such that $\overline{AC} = \overline{CD} = \overline{DB} = 1/3$. Upon $\overline{CD}$ as a base, erect an equilateral triangle CED. The basic Koch curve is then ACEDB. The ratio of length AC + CE + ED + DB to AB is the length amplification of AB. It can be shown that the amplification is

$$\epsilon^{1-D} ,$$

where ϵ is the unit length $\overline{AB}$ and D is the fractal dimension. Moreover, D can be can be shown to be 1.2618. Fractal dimensions, in general, are non-integer.

The Koch curve can be refined as shown in Fig. 1b. The refined curve, using the basic construction scheme described above, is AFGHCIJKELMNDOPQB. Note that the fractal dimension D is preserved and will continue to be as successive refinements are made. The length amplification grows as $(4/3)^n$, where n, an integer, is associated with the (n-1) refinement. Thus the length is more than quadrupled after the 5th refinement, and is increased nearly tenfold after the 8th.

SURFACE PROFILE AS A FRACTAL CURVE

As indicated in the Introduction, for simplicity, we restrict ourselves to the

Euclidean two-dimension. We know that Mandelbrot has used computer to generate Euclidean three-dimensional topographic maps [7].

The purpose of the paper is twofold: (1) to demonstrate, with the aid of actual surface profiles, the feasibility of representing the profiles as fractals; (2) to apply such models as Koch curves in predicting certain characteristics of surfaces, inferred through the profiles, which are important to tribology.

Koch Curves with Random Features -- The power of fractal geometry is that it can, theoretically, fill the gap which Euclidean geometry cannot cope with easily. The manifestation is the infinite number of fractional values of D. For the present purpose, we do not claim to have found the appropriate fractal D and its attributes for a given set of surface profiles. Our limited objective is to use simple, basic fractals such as the Koch curves. Because of its simplicity, a Koch curve in the deterministic sense will not yield a curve which looks like a typical surface profile. How, then, can we expect a feasible representation?

The answer may lie in another powerful aspect of fractal geometry. That is, fractal goes well with the probablistic character of nature. Indeed, we noted earlier that the most useful fractals involved chance. Fig. 3 consists of ten segments of straight line as shown in Fig. 1a, each of which has been given the Koch construction. Only in this case each step has two choices, i.e. equilateral triangle can be on one or the other side of $\overline{CD}$ as in Fig. 1a; and these choices are made on a random basis. This curve represents two refinements. Fig. 4 consists of ten segments of straight line and each has been given the modified Koch construction, see Fig. 2. Again, the curve represents two refinements and random choice was used in the construction; α was set at 14°29′. In this case D = 1.0488.

On Surface Profile Characteristics -- To the extent the Koch curves approximate actual surface profiles, knowledge of these models can yield useful information such as the actual length which can be used to estimate adsorption, fluid retention, for example. In the case of fluid retention, the relevance is the lubricant applied to the surface of particulate magnetic recording disks. Elsewhere in this short paper, we have referred to characteristics which involve interface deformation or thermal loadings. New insight may be revealed with such models as Koch curves or more advanced fractal surfaces.

Regarding actual surface profiles, we postulate the following relationship governing R, the ratio of contoured length as measured to the straight-line distance, and ϵ, the length of the measuring unit associated with the engineering surface:

$$R = \exp[2.3(\epsilon/\epsilon_c)^{-\alpha}] ,$$

where ϵ_c and α are parameters of the surface. Fig. 5 shows a plot of loglog R

vs. ϵ in logarithmic scale. For this set of data, $\epsilon_c = 28$ nm. The α parameter, determined empirically, is equal to 2. The value $\alpha = 2$ is quite universal, i.e. it is the case for rusted steel plate, brick, as well as plated disk surfaces. The straight line represents the equation which tends to unity as ϵ becomes larger and larger. This limiting value certainly conforms to physical reality. On the other extreme, theoretically, R goes beyond bound as ϵ decreases indefinitely. There we need to impose a cut-off on the smallness of ϵ. The circled points represent measurements by Talysurf. The half-moon points represent calculated values of R from measurements of contoured surface in the scanning electron microscope by a special technique originally suggested by D. Thompson of IBM. To show the plausibility of the Koch curves in representing surface roughness, two points are shown as the square boxes. One is placed so as to coincide with $R = (4/3)^3$ which was the amplification of length of the curve shown in Fig. 3. Based on this position the value of ϵ is 0.046 μm = 46 nm. The unit length for the basic Koch curve is ϵ/L where L is the length for which the Koch curve was normalized. Since $\epsilon/L = (1/3)^3$ and $\epsilon = 0.046$, L is therefore 1.24 μm. The second square box represents R for the modified Koch curve shown in Fig. 4, i.e. $R = (4\sqrt{15}/15)^3$. If the Koch fractal describes the material surface, then $\epsilon = 46$ nm is the cut-off of the aforementioned scaling equation $R(\epsilon)$ as fractals are scaled by ϵ^{1-D}. In other words, the two curves intersect at $\epsilon = 46$. Similarly, if the Koch kin fractal describes the material surface, the intersection and therefore the cut-off would be 100 nm. It appears that the macroscopic, exponential scaling law would intersect a relevant microscopic, fractal scaling law. The intersection thus demarcates the macro and the micro world. Further research has to be done to elucidate this complex phenomenon.

ACKNOWLEDGEMENT

The support of the U.S. Army Research Office under Contract No. DAAL03-86-K-0076 is acknowledged. Also acknowledged are a Jacob Wallenberg Foundation Grant and the equipment provided by the National Science Foundation under Grant No. MSM-8505954. The authors wish to thank Mr. David Cogdell for providing the profilometric data.

REFERENCES

1. "Some Factors Influencing the Area-Load Characteristics of Semi-Smooth Surfaces Under 'Static' Loading," F. F. Ling, *Transactions of the American Society of Mechanical Engineers*, *80*, 1113-1120 (1958).
2. "On Asperity Distributions of Metallic Surfaces," F. F. Ling, *Journal of Applied Physics*, *29*, 1168-1174 (1958).
3. "On Model Studies of Metallic Surface Asperities," F. F. Ling and R. C. Lucek, *Journal of Applied Physics*, *30*, 1559-1563 (1959).

4. "On Temperature Transients at Sliding Interface," F. F. Ling, *Journal of Lubrication Technology, Transactions of the American Society of Mechanical Engineers*, *91*, 397-405 (1969).
5. "How Long Is the Coast of Britain? Statistical Self-Similarity and Fractional Dimension," B. B. Mandelbrot, *Science*, *155*, 636-638 (1967).
6. *The Fractal Geometry of Nature*, 3rd Edition, B. B. Mandelbrot, W. H. Freeman and Company, New York, 34-43 (1983).
7. "Stochastic Models for the Earth's Relief, the Shape and the Fractal Dimension of the Coastlines, and the Number-Area Rule for Islands," B. B. Mandelbrot, *Proceedings of the National Academy of Sciences*, *72*, 3825-3328 (1975).

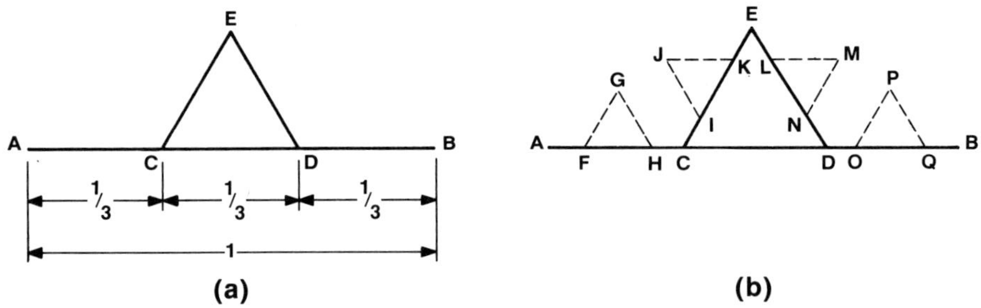

Fig. 1. Koch Curve: (a) Basic Construction; (b) Refinement of the Basic Curve.

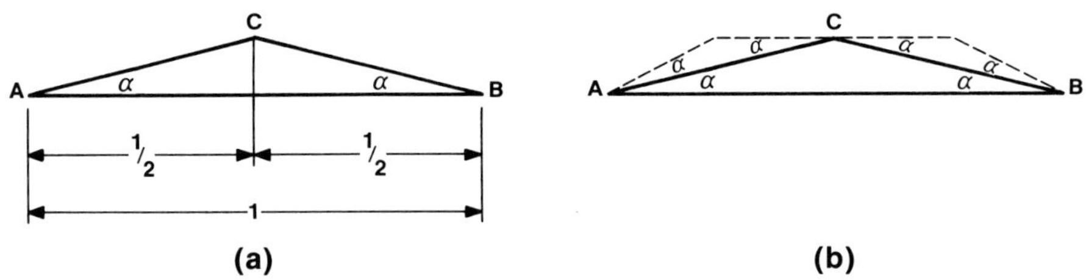

Fig. 2. Koch Kin: (a) Basic Construction; (b) Refinement of the Basic Curve.

Fig. 3. Fractal Based on the Basic Koch Curve: $R = (4/3)^3$.

Fig. 4. Fractal Based on the Basic Koch Kin: $R = (4\sqrt{15}/15)^3$.

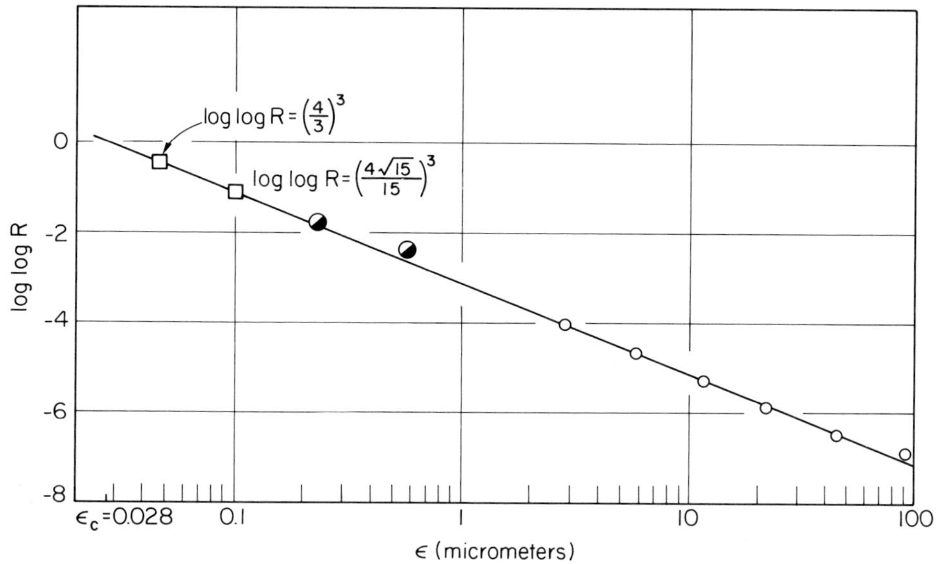

Fig. 5. Plot of Loglog R vs. ε in Logarithmic Scale.